I0473427

ART AND ADAPTATION

Art and Adaptation

A Primer from Notes

Gregory F. Tague

Bibliotekos
Brooklyn, New York

Publisher, Fredericka A. Jacks

publisher@ebibliotekos.com
www.ebibliotekos.com

Printed & Bound in the United States

ISBN: 978-0-9824819-7-4

Cover image, The Knife Grinder, Kasimir Malevich,
Yale University Art Gallery

Bibliotekos
Brooklyn, New York

"What good is a display that stays unnoticed?"
Frans de Waal, *The Age of Empathy*

"...it is not the phenomenon that is novel
but the explanation of it." Timothy D. Wilson

—

This book is dedicated to its publisher and my wife,
Fredericka Jacks, for her endless support

Table of Contents

PREFACE

Is art a free-riding frivolity with no biological or cognitive value? How, if at all, is art connected to health, pleasure, play, neural plasticity, sexual selection, sociality, and individual and group emotions? Those are big questions. While I might not have all of the answers, and since many evolutionary psychologists, biologists, anthropologists, and philosophers of art disagree, we might never know for sure if art behavior is an evolutionary adaptation.

But material culture and art making are deeply embedded in our evolutionary history. And we continue to make art. Why? Consider some universal themes across art from all ages and countries: survival, romance, family, individual values, group identity, altruism and reciprocity, religious and spiritual beliefs, and warfare. This short list is indicative of behaviors that arose from selection pressures concerning the survival of our ancestors. In terms of biology, there are striking benefits to making art over the costs, and the behavior is not only passed on by instruction and learning but the impulse is innate and heritable.

There is no single cause for our proclivity to make art. In fact, *cause* is not the appropriate word. Rather, there is a medley of adaptations that gave rise to how and why we make art. Here is a simplified sketch. Our hominin ancestors evolved a larger brain, due to a number of selection pressures, such as tool manufacture, resource sharing, and group living. The larger brain responded to survival pressures and so produced hand tools and, much later, body paints and ornaments. With the rise of larger groups and their cultures, rituals, and symbolic marks, group identity with graphic communication appeared. Perhaps most controversially, just as Darwin proposed sexual selection in many species, e.g., the peacock's tail, so too in our species art culture is a response to mating pressures.

Building on the preceding, our brain adapted to a number of pressures simultaneously, such as calculating in terms of space, objects, and events, and evolved to have intelligence modules communicate with each other. So, a bone left over from hunting could be manipulated for use, such as a tool. Perhaps the bone could be used for decoration or status, such as bodily ornament. Our evolved ability to reengineer physical objects for a number of purposes, especially associating them with our beliefs, values, and practices, gave rise to symbolic culture.

This book is not intended to be a monograph and does not mount its own original argument. Nonetheless, my fingerprints are all over it, and clearly the implication is that art making is an evolved adaptation. While the book is no more than what the subtitle says, I hope it provides both insight and entertainment to readers. The purpose of the book is to present as best as I can the views of others on the subject of the adaptive function of art.

A much revised, expanded, updated, and illustrated version of this primer appeared as *Art and Adaptability: Consciousness and Cognitive Culture* (Brill|Rodopi, 2018).

Publishing often in what I broadly term as evolutionary studies, my particular views about how moral sensations and emotions from an evolutionary perspective underlie much artistic culture appear in my comprehensive books *Making Mind: Moral Sense and Consciousness* (Rodopi, 2014) and *Evolution and Human Culture* (Brill|Rodopi, 2016). For my evolutionary writings on animal and environmental ethics, see *An Ape Ethic and the Question of Personhood* (Lexington Books, 2020), *The Vegan Evolution* (Routledge, 2022), and *Forest Sovereignty: Wildlife Sustainability and Ethics* (Peter Lang, 2025).

The Long Pleistocene
2.5mya – 11.5kya

[Simplified Graphic with Approximate Dates]

Sedentism and agriculture, 11-8kya

Upper Paleolithic/Late Stone Age 36kya-10kya

32-26kya early Aurignacian

Anatomically Modern Humans in Europe, 40kya

Global diaspora of Anatomically Modern Humans, 70kya

Earliest Anatomically Modern Human fossils, Africa, 200kya

Middle Paleolithic/Middle Stone Age/Mousterain, 300-30kya

Global diaspora, *Homo erectus*, 1.7mya

Acheulean Industry, 1.7mya-100kya, hand axe

Oldowan stone tools, 2.6-1.8mya
  ‖-*Homo habilis*, 2.5-1.5mya

Chapter One

CULTURE

Culture is a broad term that includes any number of values, beliefs, and practices of a group or across groups. There are many groups, each with its distinct culture. At the same time, I can use the terminology human culture since there are universal practices in many global communities. For our purposes, art behavior holds meaning for a particular group, but the fact is that universally human beings express meaning symbolically. A discussion of culture is necessary since it encompasses the evolution of mind, the creation of material goods, and art making.

In his *History of Art*, H.W. Janson says that because of its "intrinsic value" we often set art apart from our routine lives, whether in caves or museums. This is a modern and Eurocentric viewpoint. It is a viewpoint to be contrasted with the opinions of Andrew Shryock and Daniel Lord Smail who argue for a full study of deep history or *prehistory* among disciplines. In prehistory art behavior was closer to material culture and had functional value. In *The Artful Species*, Stephen Davies also acknowledges how early art culture served practical purposes.

Moreover, Robert Bednarik says that outside of Europe, Pleistocene art, especially of Asia, is almost ignored. Art did not originate in Europe. Bednarik adds that such material culture worked as exograms or symbolic devices for remembering. What we now call art was body paint or adornment for social status or group identity. Janson acknowledges human neurobiological universals are shaped by culture, but his notion of prehistory dates back only to about 35kya, and he mistakenly identifies early artists as

1

shamans. Nevertheless, art is a skill, as even Darwin knew, and Janson and others are therefore correct to emphasize the *making* of art.

But even here Janson borrows from Michelangelo's notion of freeing form from marble to emphasize god-like creativity more than human making. Janson's history is more to fine art, and so he lays great emphasis on meaning, "ideal form," and order. The adaptive functions of art certainly take into account how lines are cerebral and how color is sensual, seen for example in male animals that attract female mental and emotional capacities. For our purposes, the highly stylized art Janson speaks of is a very late development in a long history that has many precedents and roots in the natural world and natural processes.

With reference to Henri Bergson's *Creative Evolution*, evolutionary psychologists Leda Cosmides and John Tooby say that there is no *élan vital*, no soul and no special ingredient to what makes human intelligence. Rather, there are only genes and their effect on cell processes. The primate mind has evolved to solve adaptive problems, such as mate selection, communication, group relationships, and cooperation. Such fundamental mechanisms are universally cemented in the human mind but vary and are modified in different populations. Culture is made variously by different groups despite essentials in human nature. While there is some debate about the recognition of cross-cultural, universal emotions, for the most part there is agreement that facial muscles have evolved to correspond to emotions in any given culture. Human beings across the globe do share, however they are expressed, certain emotions, such as sadness, happiness, surprise, anger, fear, and disgust.

According to Michael Alvard, culture is the manifestation of the human transmission of inherited brain matter, i.e., neuronal information. Art culture lets us experience something we've not yet been exposed to. Culture is not merely copying but, instead, a process whereby individuals recycle such information differently and progressively. Other non-human animals copy culture, and young chimpanzees in the right environment can learn on their own. Some might anthropocentrically believe that only we have adapted a cumulative culture; we don't just copy, we understand meaning and implications. This can be true of other species, too.

On the one hand Alvard goes on to say that an important premise about our densely layered culture concerns theory of mind, a capacity to see others as we see ourselves, responsible. Mental mind reading is part of our human evolution and involved in the arts. We guess at intentions and meanings socially among people and even with symbolic objects. But Alvard insists that arts are an exaptation used for something other than what is dictated by natural selection. He claims creative enterprise to enhance social engagement, communication, or ritual is not adaptive. Stephen Davies, too, says that while art behavior is clearly tied to evolution it is not an adaptation but a mental byproduct.

Theory of mind is an important primate adaptation critical to this discussion. We need to read other minds for many reasons, ranging from selecting mates to negotiating social situations. Cultural representations help satisfy our desire to read other minds. Moreover, we expect others to attempt to read our thoughts, and so evolution has endowed our bodies with ways to express mental attributes via facial expression and especially through our eyes. Often we are only guessing at another's mind, but as Lisa Zunshine says, we do so hungrily and continuously.

Culture and the Adapted Mind

Just as genes respond to environment, the mind has responded to the environment of other people. This is true, too, in apes, especially chimpanzees, or what Jane Goodall refers to as patterns of behavior. According to Cosmides and Tooby in "The Psychological Foundations of Culture," there could be any number of selection pressures that helped evolve the mind: 1. an adaptive target, such as what to eat, whom to mate with, how much parental care to be given; 2. background conditions, such as eyes, face, light, what one needs to know; 3. design, such as environmental cues or information, causality between representation and behavior; 4. performance examination, such as how things work from the environment to the mind; 5. performance evaluation, such as how well or not the design meets adaptive target.

So a question is: Are art behaviors adaptations to our ancestral environment or spin-off responses to other adaptive behaviors? Most likely our brains evolved an adapted mind with many communicating intelligences, from linguistic to personal, in order to make art precisely for adaptive functions, such as sociality, group identity, and mate selection. Our networked brain is a variorum of evolved minds and those we interact with.

Selection adjusts organisms in terms of environment, and for highly intelligent primates like us, environment is more social than geographical. In spite of inherited species similarities, though, no two organisms respond similarly to the same environment. This has important considerations for the creation of and response to art. Evolution likes individuality, suggested in Darwin's three important criteria for evolution: variation, competition, and inheritance.

There is no static world. Culture is not out there waiting to shape and mold us. Instead, over a long evolutionary process imbedded in countless generations of human beings, we have created culture. While cultures vary, there is some underlying regularity. Values and beliefs differ, but it is human nature to have values and beliefs. This explains how and why we not only learn our own culture but can assimilate into another culture. In other words, the mind is not a blank slate at birth and evolution has not erased instincts in human beings.

To cite only a few examples relevant to our discussion, consider how according to Cosmides and Tooby we recognize faces and objects, estimate distance, choose mating partners, manufacture tools, circumvent conflict, possess social intelligence, and readily grasp emotions. Culture is inborn behavioral dispositions, especially patterns of behavior that can change based on individual choice, environmental modification, or group selection.

The mind already has, according to evolutionary psychologists, adapted content that is responsible for filling the world, not the other way around. In many respects, then, art is a manifestation of our evolutionary adaptations that with our higher intelligence fill our emotionally communicative social world. An adaptation ultimately deals with successful reproduction of healthy offspring but is allied to survival for mating. Physical development and parenting are for sexual maturation. Once fully developed, mates can attract each other and reproduce. In many mammals, the shape and color of the physical body signals health and fitness. Even among human beings, studies demonstrate how we respond positively, visually and emotionally, to attractive mates.

We don't necessarily learn art from some socially imposed culture. Culture is by definition restrictive, so certainly it influences and shapes inborn art behaviors. And we can, of course, learn how to make art depending on the culture we are in. But notice how easily everyone as a child makes art, and then as an adult responds emotionally and intelligently to art. Even a negative response to an artwork is an emotional and psychological reaction based in some part on our evolutionary past as well as defined cultural environment.

Such naïve skills and mature responses are inherently built into our adapted mind, not imposed from the outside. Certainly our upbringing and defined culture can affect the expression of our emotions. Contrary to Piaget's schematic of learning in stages, Cosmides and Tooby say infants are born with concepts of objects, distinctions between animate/inanimate, and spatial notions of movement and distance. Likewise, Paul Bloom, a developmental psychologist, has done compelling work demonstrating how certain emotional modules, such as caring, already exist in the mind of an infant.

Just as we are all born with genes, we are all born with mental functions, i.e., they are not acquired. These evolved functions have been born and re-born in human beings, genetically transferred and inherited, many billions of times. Cosmides and Tooby say that "All humans tend to impose on the world a common encompassing conceptual organization, made possible by universal mechanisms operating on the recurrent features of human life" (91). In other words, we can understand the art and culture of other, even ancient, people, since such material and symbolic forms are rooted in us.

Roy Baumeister seems to define our humanity, in this regard, by saying that "...Nature built us for culture" and that "we evolved to be able to take advantage of...[the] benefits of belonging to a cultural system" (ix). Borrowing from Denis Dutton, Anthony Lock has argued that we find distasteful any forged symbolic representations we've aligned ourselves to culturally. Part of our inherited human value system includes original creativity, then.

What we broadly call culture, or our values and practices, became a plan to negotiate the social and to manipulate the physical environments in terms of survival and reproduction. A key component of our evolving beliefs about other people grew from both food production and sharing, first with offspring or mates, and then for close non-kin, and later, strategically, for others. In evolutionary terms such practices don't come from nowhere. An examination of work by pioneering primatologists such as Dian Fossey, Biruté Galdikas, Frans de Waal, and Jane Goodall reveals that great apes, too, exhibit complex social organizations revolving around families, coalitions, and groups.

We don't need large brains to get food. However, we do need large brains, and so have evolved a large brain, for complex social interactions. This is Robin Dunbar's social brain hypothesis. Baumeister seems strongly to say we have survived and dominated not merely by social intelligence, but in how we have shared socially and culturally such intelligence. There are other social animals, says Baumeister, but only the social/cultural one, the human being, has been able to build vast and different societies from shared informational knowledge and symbolic behaviors. This means that beyond the physical or emotional connection among individuals we might see in great apes and other primates, we have become part of and intimately linked into our cultural networks. For example, we have the ability to translate ideas from one culture or language to another.

Baumeister suggests that part of any organism's evolutionary suite of adaptations would include some ability to comprehend its environment. For highly social human beings, the need for understanding, to estimate and to plan, evolved even more complex mechanisms for control. A simple organism maneuvers around in its given environment. But a human being has the ability to choose and control an environment, including the environment of other people. Part of the understanding and control of such environment would include artistic culture.

Gene/Culture Co-evolution

Peter Richerson and Robert Boyd write about the co-evolution of genes and culture, with stress on the latter. Our values, beliefs, and practices, while inherently biological, arise from culture. Culture affects survival, so some cultural variations persist while others do not, say these authors. An environment of genetic adaptedness is established with the rise, spread, transfer, and diminution of cultural practices. Citing Darwin and his observation that species are gene pools, Richerson and Boyd importantly suggest that culture evolved because it is more capable than genes in quickly spreading values, beliefs, and practices. Nonetheless, individuals can affect a population. Fads, fashions, and celebrity trends satisfy an immediate need for attention, to fit in, and a deeper need to belong to a group and cooperate.

Variation that affects behavior, whether good or bad, to such an extent that it is noticed and copied by others, will account for diffusion. While we need genes to imitate, learn, and remember, such rapid and widespread dispersal of behaviors that can be copied is what we call culture. There is simply far too much for one individual to learn all on her own, so cultural transmission accounts for much of what many people do. Notably, there is consensus recognizing culture among apes, too. Nevertheless, most culturally communicated information does not operate like a gene and can be transmitted incorrectly, since many cultural behaviors are essentially cognitive.

Our culturally active brains have been evolving for over two million years. Culture must hold some evolutionary adaptive function. Otherwise, it would have been selected out of those innate parts of our brain, such as emotions and instinctual dispositions, which enable it to happen. We imitate, a form of free-riding, because it is easier than reinventing the wheel. When we imitate we improvise and improve, and in this way culture is cumulative, akin to descent with modification, and can change rapidly and widely. No other species is quite like us in this regard, but there is study of cultural practices and transmission among chimpanzees and other primates in terms of tool selection, tool use, and other practices such as food handling.

Richerson and Boyd disagree with Cosmides and Tooby who see culture as *evoked* from an *innate* psychology, such as emotions related to mate selection and manifested differently because of environmental cues. Eckart Voland and Karl Grammer go as far as saying that our ability to detect beauty is innate and demonstrated in infancy. We have a mental as well as a physical reaction to beauty. Rather than emphasizing innate universals that aid in learning, Richerson and Boyd stress the adaptability of cultural transmission.

While Richerson and Boyd emphasize culture over genes, such a position does not disregard the biology of culture or the adaptability of art. Nature-nurture debates carry little weight among biologists. While both genes and environment contribute to the survival of an organism, one could emphasize one side of the equation over the other. Here are two simple examples of gene-culture co-evolution that would impact this discussion. The development of better hunting techniques reduced a robust body, and hence decreased dimorphism and generated more cooperative sociality. Bipedalism and simple speech eventually favored the lowering of the larynx for more sophisticated language, and hence eventual symbolic communication.

What has probably happened is that we have constructed on top of the small group mentality that developed during the long Pleistocene of our hunter-gatherer forebears, approximately 2.5mya to 11.7kya, a social psychology that permits us to function on a large scale, say Richerson and Boyd. By analogy, we have two brains: on top of the older so-called mammalian brain is our more mature neocortex. This view takes into account the importance of social intelligence and then symbolic and graphic communication.

Human beings can be more cooperative and altruistic than other species, and we have far more cultural transmission. So our cooperative tendencies are not just genetic but have spread because of culture. According to Richerson and Boyd, genes have responded to and have supported cultural adaptations. But as a species, of course, we are more different from other groups culturally rather than genetically. We share about 98% of chimpanzee genes, but we are not chimps. While natural selection primarily works on the individual, not the group, clearly our genes have benefited from cultural forms of adaptation.

Culture and Social Selection

Alex Mesoudi says that any explanation of human behavior must include the cultural and biological, and so he relies on a Darwinian template. Culture is not a mere backdrop that affects human behavior but rather is constructed by and is malleable to human needs, desires, and emotions. As a psychologist, compared with the biologist Mark Pagel who will be discussed, Mesoudi leans quite a bit on social constructionism in spite of his evolutionary bent. While he argues for a Darwinian model of culture (variation, competition, and inheritance), he nevertheless makes culture the defining force on an individual almost to the exclusion of discrete genes, but this is not to say he completely ignores genes.

Mesoudi says that genes enable certain behavioral capabilities to flourish under cultural conditions. Genes themselves are not responsible. Surely no single gene is responsible for any single behavior. Rather, genes are either turned on or not, and work in concert with other genes. At any rate, similar to Mesoudi, one finds Richerson and Boyd. For Mesoudi, culture is data assimilated from others and "stored in the brain as patterns of neural connections..." (3). Simi-

larly, Eva Jablonka and Marion Lamb discuss not only genetic but also the behavioral, epigenetic, and symbolic dimensions of evolution.

Culture is information that affects behavior, not unlike
genes. But anyone's behavior can be the result of genes,
culture, or individual learning. For example, to cite a perhaps familiar distinction, Western (especially American)
people emphasize individual responsibility, more at analytic
thinking, while East Asian people will consider a whole
situation, more at holistic thinking. So in this case, behavioral variation is due to cultural learning. Nicholas Rule
says that even theory of mind has been shown to be culturally dependent, since each ethnicity supposedly reads the
eyes of its own better. Like Richerson and Boyd, Mesoudi
says that genes are responsible only for a learning capacity
and not for specific cultural contents. In fact, Mesoudi goes
on to say that in many societies people do not learn individually but through groups, such as the Amish.

Cultural evolution is not ladder-like and not progressive,
as per Herbert Spencer, the one who coined the expression
survival of the fittest and who promoted Social Darwinism.
Societies do not progress up an incline. Instead, there are
variants within a population that through natural selection
(variation, competition, and inheritance) make change over
time. Biological inheritance is on the micro-level, i.e., selection, mutation, drift, and is not, on the surface, blending of
traits. If evolution were only a blending of genes and not
transference, traits would be blended out. We get only one
version of any gene, not a blend of it, e.g., eye color. Cultural traits, however, can blend or not, says Mesoudi, because on a neural level culture is discrete.

Mesoudi's emphasis then is on learning. Since guided variation is individualized learning, there is much diversity in
the finished product. With prestige learning, on the other
hand, there is more consistency in the finished product since

many people copy the model from someone who has had the most success. Over time, it is possible, says Mesoudi, that diversity will level off and disappear if everyone learns how to make the best, i.e., prestige product. Nevertheless, there might not be one and only one prestige product, and all of this may depend on the adaptive environment.

For example, Richerson and Boyd talk about content bias, or how attractive any cultural product or idea is and the likelihood of it getting copied. They say that disgust stories spread widely, probably because from an evolutionary standpoint we have been concerned about infection and contamination. As well, stories of the supernatural where some creature has super powers but resembles a human being are also cross-culturally prevalent. Evolutionarily, we tend to conform and not deviate, and that is how groups are formed and transmit values, beliefs, and practices.

Mesoudi says, then, that there is *cultural group selection* which would have found certain genetic tendencies advantageous, such as cooperation and altruism from the mother-infant bond. We can see that culture fits this model, for it selects out of any group those traits which do not in some way benefit the group. For our purposes, artistic behaviors, whether cave painting by young men or ritualized tattooing, spread as a means of group inclusiveness.

Culture and Epigenetics

At the beginning of his book, *Wired for Culture*, biologist Mark Pagel, offers the astounding statistic that, at the time of his writing, there were about 7,000 languages across the globe, some within a stone's throw of each other. This fact alone is evidence of the highly-fractured tribal mentality of culture that is part of our species. In sharp contrast to some animals of the same species, we often cannot communicate with other human beings in spite of our linguistic capacities.

In terms of cultural evolution and transmission, however, ideas themselves can, as Pagel says, "jump from mind to mind..." (3) in an epigenetic fashion.

The epigenome is DNA-like compounds in the environment that can turn genes on/off. These compounds derive from food and chemicals in the environment, whether naturally occurring or manufactured. This means that one's epigenome can change dramatically from time to time.

In other words, epigenetics explains the explosion of so many cultures and cultural practices, even between different cultures which might copy only certain behaviors. In spite of our parentally inherited genes and social environment, we are able to explore any ideas from any culture. Indeed, as a species, we tend to use the accrued information found in culture more, to some extent, than our genes in addressing the problems of existence.

Proof of this cultural ability to choose is that a human baby can be brought up and thrive in any culture, but a wolf reared by sheep is still a wolf, says Pagel. We don't adapt to any environment, per se. Rather, our humanity allows us to adapt to the environment of other people and their cultural practices. We still have an evolutionary disposition to survive, but like a biological parasite, a cultural meme (Richard Dawkins' term to characterize a cultural unit) can affect us whether we want it to or not. Such copying behaviors include degrading one's health or following an ideology that puts us in harm's way.

Anatomically Modern Humans (AMHs) circa 200kya were distinct from other *Homo* species, such as Neanderthals and *Homo heidelbergensis*, as well as pre-modern people, such as *Homo ergaster* and *Homo erectus*. Forms of culture early people possessed seem suddenly to have bloomed and flourished. Whereas stone tools existed 2.5mya, by the time of AMHs we have ritualized symbolic culture and artifacts. Andrew Shryock, Thomas Trautmann, and Clive Gamble note that artifacts exhibit cognition through material objects. We are a thinking species whose ideas are implicit in the material objects we make. There are, as examples, pierced and painted shells and engraved stone in Blombos cave South Africa circa 75kya, as well as beads of red ochre probably to paint something, from Pinnacle Point South Africa circa 160kya.

Scholars debate the origins of symbolic culture. More close to 80kya we begin to see, says Pagel, jewelry like shell beads, teeth, ivory, or ostrich shells strung together; bone, antler, and ivory artifacts; bows and nets; more sophisticated tools. Before shells were drilled so as to be strung together, they might have been collected and shared nonetheless. McBrearty and Brooks, followed by April Nowell, place symbolic art forms much earlier and even among Neanderthals. Robert Bednarik says bodily adornment, or self-awareness stitched across hominids and hominins alike, dates to 2mya and renders any cognitive super development at 50kya or thereabouts as unlikely.

As Michael Tomasello says, we are one of the few species that does not simply mimic observed behavior in our learning. We improve what we learn, improvise, and make our productions more sophisticated, which in turn are imitated and improved upon over and again. Part of this stems from our theory of mind which enables us to take on another person's point of view. This perspective-taking is important socially since it helps us attempt to comprehend the significance and meaning of other people's actions. We re-tool mental behaviors and value-laden attitudes.

Mind Sharing

Theory of mind is important in terms of social learning, and so culture is more mind than genes. Of course genes build minds. Theory of mind works for artists as well as viewers. Visual art is meant to be seen, and therefore the creator has, even subconsciously, an ideal viewer in mind. Even so, the artist is aware that speculative guessing about intention, meaning, content, and form will occur in any viewer. Art culture is social mind guessing and sharing, a set of cognitive more than merely instrumental adaptations.

In biology there is selection that favors more than one outcome. Culture permits this multiplicity and generates a type of specialization of talents, a sorting, that also requires co-operation. For instance, one can build a hang glider but not necessarily have the dexterity to fly it. According to Pagel, there are broad categories of culture people sort themselves into because of traits: 1. aesthetic (creative culture); 2. cerebral (information); 3. communal (relationships and emotions); dark (intense and hedonistic pursuits); 4. thrill seekers. In other words, culture differentiates human abilities. At birth, gene scrambling permits some individuals to succeed at tasks others are capable of appreciating but unable to perform, and such diversity of talents has helped us to survive.

Pagel suggests that cave painting skills did not spontaneously arise but were adapted for much earlier and seem to appear suddenly only because of favorable conditions. We will talk about cave paintings later, but for now we can say that in the Ardèche region of France, Chauvet-Pont d'Arc Cave, there are depictions of rhinos, mammoths, panthers, bears, horses, and owls. There is an eight-legged bison suggesting movement. There are multiple images of rhinos in various sizes. There is the hand of a young man outlined with paint, from approximately 36kya. These paintings could have been the result of any number of prompts, such as to catalogue species for hunting, to note environmental information, to create a cognitive map, or to illustrate part of some ritual. Bottom line, though, is that art represents an individual's mind externalized. At times the individual mind physically represented might characterize group thinking.

At other times the individual mind might oppose group thought.

Interestingly, there is no real evidence, says Pagel, that later paintings are the climax of earlier, weaker predecessors. Paintings from Lascaux Cave at 18kya do not exhibit any further sophistication than the paintings twice their age.

According to Pagel, the "arts...evolved to enhance the expression of our social behaviors" as "cultural enhancers" related to emotions (132, 135). Art culture magnifies human expressions that are species inborn or individually latent. Art can be a sexual signal, such as: Look at how much energy I can waste doing this and still survive, so why not reproduce with me? At the same time art is a sophisticated symbolic mechanism that ushers forth values, beliefs, and ideas. The human brain is more complex and has developed much faster in evolutionary time than any other, which is perhaps why we survived over other hominin species, and part of this explosive development can be attributed to physical and mental communication through cultural productions such as art.

Chapter Two

ADAPTIVE FUNCTIONS

In his article "Imagined Worlds," literary Darwinist Joseph Carroll outlines some functions of art that could be adaptive. For instance, pictorial representations could supply geographical, climatic, and wildlife information. Visual art could be a means of offering a different solution to a problem. Plastic and musical arts could provide value symbols to bind members of a group. Any form of art could have served to stimulate cognitive mechanisms to enhance memory, engage in mental play, or permit display of social status or sexual desire. The bottom line, says Carroll, is that whichever of these functions one takes into account, all engage the human imagination, which is a mentality to perceive, process, contemplate, copy, and respond to the social and natural worlds.

Terrence Deacon, in *The Symbolic Species*, says our artistic creations were not necessarily inevitable and were perhaps products or byproducts in reaction to changes of the physical geography, climate, and pliable materials. Deacon claims that there is no solid connection between artifacts and enhanced communication. The question about whether or not the arts are evolved adaptations remains, but the position taken here is that arts are adaptive. Certainly because of runaway culture some forms of what we call art, such as fads or fashion trends, serve no adaptive function now. The larger the community the more varied the art cultures, where some can even be self-debilitating.

At base we have art since it is part of our neurobiology at play, stems from our symbolic and imaginative capacities, helps some people distinguish themselves from others, and evolved from means to attract and select mates. After all is said and done, it might simply be that the human mind and its cognitive functions are the adaptations from which art behaviors derive. But that approach seems too broad and skirts the subtle questions addressed here. Some suggest that academics like me push for an adaptive function of the arts because it helps justify our scholarly endeavors. That's a hard pill to swallow.

Adaptive Problems and Questions

There are questions concerning how art could have benefited an individual or a group in terms of survival and reproduction. What were the advantages gained by making art, a procedure that involved the cost of energy in finding the correct materials and then manipulating them. As E.O Wilson might say, from *Consilience*, a simple answer has something to do with the fact that but for us, all animals are governed by instincts. We, on the other hand, have evolved extraordinary mental capacities that have given rise to a number of traits including the making of art.

Art culture is like offspring in that costs are expended since benefits will be gained. Advantages include, primarily and for many reasons, the attraction of another's attention. Costs should be lower than benefits. At least in prehistory, some of the advantages gained in the rise of an artistic adaptation would include, for example, better cooperation, rise in status, or enhanced social standing in the community. Mental play could be seen as a byproduct of some of the other adaptive functions. Since there is a risk element in making art, that risk might have solved the problem of low status or competition among early men. This is not to say that early women did not compete in body decoration to attract a number of potential mates from which to choose.

Artistic culture is predominantly an evolved psychological mechanism. Sophisticated behaviors flow from cognition.

Our advanced intelligence, going back perhaps 2.5mya when stone tools first appeared, evinces a cognitive and behavioral flexibility not seen in other organisms. Clearly then, as is true today, artistic productions, whether an advanced stone tool or cave painting, demonstrate not only capacities for mental representation but the ability to plan and execute an artifact out of such. What would be the adaptive value of this type of production? This is not to say there are specific genes for art. Rather, there are genes for creativity and imagination in all human beings, and once creativity began, what we now call culture helped spread the practice.

On the one hand, such productions differentiated individuals, so there might have been some aspect of sexual selection, as Darwin saw in the colors and antics of birds. For instance, if I can expend the time and energy to fashion this beautiful stone tool, which I might never use, then I am more fit, mentally and physically, than those other guys who just sit there doing nothing. So, mate with me. Technically, fitness equates to healthy offspring who survive to reproduce. On the other hand, the tools could have been traded for favors or resources. Even male chimpanzees will trade meat for sexual favors or enhanced reputation. For early people, at the same time, such tools might simply have been a means to distinguish high status, for whoever could make such a fine product should be a leader.

Finally, at a later date, wood and stone artifacts, beads, tattoos, skin piercings, color pigments and such could have been part of ritual performances. Beads, for instance, of about 30kya were probably more for social status than mere personal decoration, says Steven Mithen in *The Prehistory of the Mind*. Beads and pendants, however, have an older tradition. Initially, colored pebbles, bone, or teeth were collected, shared, and displayed. Why? Was there absolutely no fitness-enhancing component, such as direct/indirect reciprocity, to the practice of gifting or sharing distinctive items? As technology increased, these ornaments could be shaped and drilled for personal adornment. Importantly, such ornaments held not only ritualistic value but personal, symbolic, or cultural attitudes.

Coming back to inventive mental functions in terms of the co-evolution of genes and culture, there obviously was some adaptive value to imagining another realm. The creation of another representational dimension has been planted into us, no matter which world culture, from an early age and has persisted for millennia. Children from the earliest age, without provocation or training, make both realistic and imaginative art. Nearly everyone listens to music. Adults engage in either the making or appreciation of various forms of art, whether writing or reading a novel, directing or watching a movie. Art pleases us, and if it did not, nature would have selected it out of our repertoire of functions.

Correction and elaboration: art does not just please us cognitively, but it fosters sociability through graphic communication. It can also demarcate boundaries between groups. It also implies an attentive participant. Those are fitness-enhancing traits. For example, the team of Pierre-Jean Texier et al. examined 270 ostrich shell fragments from at least 60kya. They conclude that deliberate markings and patterns represented on the shells are an early yet advanced system of intentional graphic communication for social and cultural purposes. Such innovative symbolic printing did not erupt overnight but developed incrementally through time from earlier forms and practices.

Artistic creation and admiration is not an added function, not entertainment, not enrichment, and certainly not entitlement. While it might not be labelled art, Jane Goodall, in her book *The Chimpanzees of Gombe*, notes how some of these intelligent creatures in their natural habitat have engaged in what can only be described as doodling. Wolfgang Köhler quite often witnessed many chimpanzees displaying on their bodies bits of rag, rope, twigs, or even metal to adorn themselves. Köhler is convinced that this behavior rises above play and indicates a heightened self-consciousness in order to attract attention. Robert Bednarik says chimpanzee self-awareness is indicated by how they decorate themselves. Large brained species with extensive social networks self-decorate.

There is no accident for ape/human continuities. Consider the title of one of Frans de Waal's most popular books, *Chimpanzee Politics*. So too with us: the arts imply an active association with individuals or a group. Much art behavior, while individually manufactured, serves a social function, whether for inclusion or exclusion. The ability to make and the capacity to appreciate arts are built into us at conception. Culture is our ecosystem of art behavior, and we measure each other culturally.

Darwin and Natural Selection

At this point it might be instructive to look at Darwin's definition of natural selection, which emphasizes competition among animals where those who survive possess advantageous variations. Such competition is based on fitness-enhancing mechanisms, but the difference for us is that we compete in terms of mental abilities. Genes respond to environment, but because human beings for time before recorded history have mastered physical environments, we've had free reign to select certain features and mental capacities.

This is not to say animals do not possess mentality. In fact, researchers such as Darwin, Dorothy Cheney, Richard Seyfarth, Jane Goodall, Sue Taylor Parker, Kathleen Rita Gibson, and Frans de Waal, to name a few, have convincingly demonstrated mental capabilities in monkeys and mental abilities in apes. While there are intuitions and instincts in human beings, they are quite refined. Our intelligence and reason are adapted functions that work over instincts. This is not to minimize instincts since, for example, we can be saved by fear. There is memory, attention, and imitation in human beings and other animals, although we have imagination and what evolutionists now cite as our symbolic and abstract capacities.

In *The Descent of Man*, Darwin, however, insists that animals have a form of imagination and notes, for example, a dog howling at the moon. Darwin also attributes a form of reason to animals, giving the example of pack dogs on ice who separate to spread weight, or a dog looking for water who knows to move to lower ground. Earlier than Darwin the German philosopher Arthur Schopenhauer argued for understanding in animals, citing the elephant who would not cross a rope bridge knowing its weight would spell disaster.

Some might argue these are instincts, but less so, says Darwin, with the example of the elephant blowing through his trunk to move an object toward itself or a bear using its paw to make a current in water to bring a piece of bread closer. How does the saying go? If you collect enough anecdotes, then you have data. Furthermore, Darwin tries to make the argument for some type of abstract thought or self-consciousness in animals. For instance, he writes about a dog who knows the concept of another dog, and then offers us the example of a dog dreaming.

All of this is relevant since, in spite of our advanced capacities, we are related elementally to other organisms. There is a continuum among species, as Darwin's branching diagram and subsequent versions of that model have proved. Our humanities and arts, although by now driven almost entirely by culture, have their roots in natural functions that helped us survive and reproduce. Natural selection is a sorting process to achieve a functional composite. In no way is natural selection either teleological or evaluative. For our discussion this is important since all things human, from emotions to facial features, are the results (composites) of a long process of sorting to find what works best.

Evolution builds on what is already there. We are highly evolved, but not exceptional. We see elemental correlations of our physiology and behaviors in other primates, especially great apes. We share facial expressions, gestures, vocalizations such as laughter, the need for individual status in a group, sociality, coalition formation, as well as emotions ranging from fear, caring, to anger.

Darwin and Sexual Selection

In Darwin's estimation, many of the mental faculties were evolved adaptations concerning discriminations for sexual selection, or how one individual has an advantage over another of the same sex and species in terms of reproduction. Female birds will choose among many males, and most likely choose the most fit. In this way, males will advertise their fitness with elaborate song and brilliant colors to demonstrate how they can expend considerable time and energy and yet survive well. One instance is the peacock's tail, where he exerts considerable effort in charming a female.

Biologist Richard Prum says that types of biodiverse art display co-evolve along with the evaluation of such display. There is cognitive feedback so that new forms of display can emerge, such as a variety of bird songs across generations. Considering the continuities between animal and human species, such a theory has immense implications in terms of art behaviors. Art making culture breeds, borrowing an expression from Stephen Davies, an artful species.

Darwin draws connections between the animal and the human worlds. Therefore, some of the evolutionary forces that have shaped them have affected us, too. For our purposes, as per Darwin, what we call art has roots in the natural, biological world of other creatures and our own deep prehistory. We can't ignore the fact, though it is easy to do so for millions of people nowadays, that in most of our evolutionary history we lived in close community with a variety of varieties of plants and animals. Their shapes, movements, colors, textures, fragrances, and sounds are ingrained in our adapted behaviors.

Art is not something people make only because there is nothing else to do. There are biological imperatives behind art, although, now, in our highly developed world, much art might be in response to cultural cues. For some people today, art is only business and not creativity. Biological imperatives stem from natural and sexual selection. That is, stone tools helped us survive better by facilitating butchering for meat proteins to build our brains. Decorative arts, such as tattoos and piercings, provided a means of social and sexual selection. Such material arts serve a function through their form.

Beauty and the sense of color are not limited to human beings, says Darwin. Apes like the color of ripe fruit, an adaptive function. In many cases (for jelly fish and lizards) color derives from natural selection to make them appear poisonous or unpalatable to predators. In terms of sexual selection, color is attractive to females, e.g., certain dragon flies are attracted by color. Not coincidentally, there is no color in blind beetles. Some butterflies and moths have colored wings, but only on one side for strategic display. Darwin was convinced that in nature much of the color on male creatures, usually the most colorful and ornamented, from insects and fish to mammals, is for exhibition and performance. Such conspicuous ostentation functions to attract females of the species and demonstrates the mental faculties of "discrimination and taste" (*DM* 465), says Darwin.

Some characteristics of color or other ornaments, e.g., horns, have been transferred to females, but the male typically has more color and more pronounced features. Male mammals differ from birds in that they battle with teeth, tusks, and horns, whereas natural selection eliminated war characteristics in females. At the same time, such huge, branching antlers are purely ornamental. Darwin offers many examples of how such horns used by males in battle are only effective against others with the same, i.e. a result of sexual selection in competition for the female of the species. Antlers and horns, when too large, are a handicap, so, according to Amotz Zahavi, a means for a male to advertise his fitness. If he can carry around such large antlers that are useful only to battle another male of the same species, he must be fit enough to produce healthy offspring and provide for them.

We don't have large canines or antlers, so the suggestion by some authors, such as Geoffrey Miller and Denis Dutton, is that art making serves the same function as sexual selection. I will discuss these authors shortly.

The selection for these features of color, song, and morphology implies some type of mental faculty of choice in a female of the same species. None of this, however, precludes male choice, though generally the male will mate readily and often indiscriminately. Males advertise and females choose, especially true of birds. In fact, note how to this day human females apply bird feathers to their apparel. Victorian women used to position flamboyantly colorful feathers in their hair. Darwin notes that in some cases fights between males are not necessarily competitive per se but provide a means of physical demonstration when there is a female witness.

In *The Descent of Man* Darwin spends considerable time discussing ocelli, or eyespot designs, on the feathers of male birds. Darwin is convinced that an ocellus, effective only at a certain angle, is an adapted feature via sexual selection and evolved from spots and lines. The ocellus attracts attention. Surely, why are we so attracted to and awed by the peacock's display? Male birds with brightly colored and large feathers seem aware of their own beauty and are sedulous about pruning, and such birds are more quarrelsome with other males during mating season, says Darwin.

Curiously, though, the beautiful feathers we now see are an example of exaptation. Originally, they evolved to maintain heat and became primarily useful in flight, later for nesting, water repellent and floating, and finally for sexual display.

If color and ornament are for protection and defense, why, especially in birds, is it so pronounced in males and not females? Typically a female bird has duller color relative to the male, and probably for protection. The answer, according to Darwin, is that ocelli and colors, originally selected by females judging such males to be more fit, evolved into more elaborate designs and colors as species varied. Colors are innate and designed for attractiveness, evidenced in how the male's brightness, will increase with sexual maturity.

In *The Biological Origins of Art* Nancy Aiken spends a considerable amount of time discussing ocelli or eye spots. These features are threatening to many species and almost act like a mask to deceive. Even in our own art culture, consider how much emphasis is placed on portraiture and on eyes in portraits. We like masks and masquerade since they tend to hide or exaggerate features. With a focus on eyes, staring can be threatening across species, so it is not accidental that ocelli are prominent and prevalent.

On a simple level, Darwin believes colors and ornaments among animals and insects have parallels in human culture.

We see this easily enough in cicatrices (scarring) of flesh, which can change over time because of variable taste or exposure to other cultures. Terrence Deacon suggests that although there seems to be some essential human neurology in basic color reference, it is culture rather that makes color terminology universal.

Sounds, movements, colors, and patterns obviously give pleasure to many female species in the natural world, as well as to human beings. In order for such preferences to survive and not be selected out through evolution, there must be an adaptive advantage, which Darwin calls sexual selection. Males tend to advertise their fitness in any number of ways so that females can choose one who seems best to engender offspring. In terms of the so-called selfish genes, this is no light matter. First and foremost, species need to reproduce, and second the offspring must be able to survive so that the genes will persist well into future generations.

Near the end of *The Descent of Man* there is a section entitled "The Influence of Beauty in Determining the Marriages of Mankind." Here Darwin reminds us of our human passion for body ornament in terms of status and groups, social and cultural selection, and sexual selection as well. These are common, human threads. But tribal differences over time determine the difference in degree of practices from one culture to another. Though not widely known, about 70kya there might have been a population bottleneck with as few as ten thousand AMHs. All of our ethnicities and cultures have descended from that rather small pool.

Practices Darwin discusses, still prevalent in some form or another, include how and where to place scars, body piercings, tattoos, skull shaping, hair manipulation, teeth sharpening, and ornaments to the nose, ears, or mouth. These behaviors continue to this day in how people surgically manipulate their bodies according to perceptions of beauty.

Although such practices are evident in women, men, according to Darwin, seem more often the most ornamental.

In current culture, men tend to advertise their desirability as mates with conspicuous consumption, such as expensive cars. With our current human culture, the most powerful, prestigious and attractive men are the ones who exert some choice over women. This does not discount Darwin's ideas, which change by virtue of cultures, since human female adornment is most likely designed to attract many men so as to be able to choose the best from among them.

Considering what Darwin tells us about mate choice, Richard Prum adds that in order for aesthetics to evolve there should be two requirements. First, an individual must possess an element that another perceives and registers as a signal. Second, the perception of a signal element is cognitively evaluated leading to some type of preference for it or not. With any such discrimination for signals, an aesthetic sense can evolve. Prum leans to sexual rather than natural selection. He confirms Darwin's emphasis on the reproductive benefits involved in aesthetic preferences and emphasizes that most likely there has been a co-evolution of both the signal and the perceiver's judgment.

The physical and mental interplay among display, perception, and discernment between individuals is, then, a fundamental driver of art culture. Daniel Smail and Andrew Shryock note that historically and culturally, from Queen Elizabeth I to a Hawaiian King, ornament and exhibition represent our deep history and continuity not only with early hominins and hominids but with other species.

Sexual Selection and the Hand Axe

In terms of human productions, Steven Mithen and Marek Kohn examine hand axes and note that many of our hominin ancestors going back over 2my and subsequently from Africa, Europe, and Asia produced thousands of examples of these stone implements. Acheulean hand axes of the Lower Paleolithic date to about 1.4mya and are oversized and thus not useful, possessing far more symmetry and polish required for any practical applications. Such overly crafted implements imply that these were not simply tools for use but some form of art culture or products implicated in the sexual selection for mates. In a similar vein, Daniel Smail, Mary Stiner, and Timothy Earle say that any attractive material artifact communicated any number of sexual signals, from trust to prestige.

Mithen and Kohn see the hand axe as art and raise some important points. They ask, for instance, why there are so many hand axes. They conclude that the tools were manufactured because of sexual selection pressures and male competition for mates. They ask why such devices are abundant at certain sites and suggest that this is so early women could actually see the men making the axes with no chance of cheating. They ask why so much time was devoted to making anew many of these artifacts when re-touching others would have been as effective. The authors determine that the tools were not just practical, but in shape and design they served as repeated fitness indicators of manual dexterity, strength, and cognitive ability.

These authors go on to ask why so many of the hand axes display a high degree of symmetry. They conclude that the tools are symmetrical because we have evolved a bias toward balance and equipoise, especially in terms of sex, e.g., body and facial features. Such tool manufacture demonstrates whether or not one can accomplish this preference. Finally, how to explain the especially gigantic hand axes. Overly large tools or those of other superfluous design acted as meaningful social displays by early men to advertise fitness to women and competitiveness to other men.

In addition to its obvious utility, there was some type of social significance to such tools, and more particularly, sexual selection in a socially complex and competitive environment for mates. In terms of the theory of sexual selection, the tools functioned as indicators, much like a peacock's tale. In other words, the ability to expend time and effort to make such an attractive or large tool indicates that the maker has good genes. Also, the tools would function as aesthetic displays that were meant to call attention to the maker as opposed to any other potential mate. Modern courtship displays, Mithen and Kohn note, as per evolutionary psychologist Geoffrey Miller, would include a large vocabulary and a sense of humor. Some, to be mentioned soon, debate the reach of Miller into sexual selection theory.

Our ancestral hominin groups were fairly large and certainly socially competitive, as we see clearly in our nearest living ancestor, the chimpanzee. (The bonobo tends to be more peaceful.) Therefore, in addition to its obvious utility, the symmetrical hand axe served as a reliable indicator of fitness. Such fine craftsmanship showed an attentive woman how a prospective mate could obtain resources, i.e., perceptual skills and cognition, and how a man could plan, i.e., strength and excellent eyesight. Additional fitness indicators would include the ability to observe and so distinguish oneself from others in the group. Female members of the group probably made hand axes too, but they were no doubt more practical and less refined, perhaps for digging food resources from the ground.

Cognition and Cooperation

Michael Balter says that art is the communication of a cognitive ability to make meaningful symbols. The earliest known first occurrence of symbolic art is probably not the first occurrence, which might have been lost or not yet found. Chances are very good that early tools and other artifacts, for instance, were made of wood. Chimpanzees use wood and stone tools. No doubt the fundamental mental and even manipulative components of making symbols preceded art, with the open question of whether or not symbols helped early people. Cruciform art appears in caves to almost 40kya, and there are simple, portable etchings much older. These physical manifestations post-date older mental functions, where prehistoric people would have been attentive to the lines, patterns, and colors of their environment.

Balter sees tool making as an early, proxy behavior for the creation of symbolic forms. For example, brain scans of skilled stone knappers reveal for the manufacture of early tools visual and motor areas. Brain scans reveal for the manufacture of later, more symmetrical tools language areas as well. Symbolic art does not just suddenly appear, although that is what seems to happen in the artifactual record. Alternative forms, not necessarily art, in friable and easily decomposable elements certainly came much earlier but are lost in the sediments of time. Nevertheless, the prehistoric artifacts that remain point to individual cognition in a cooperative culture.

Stephen Lycett says that prior to Darwin, most thinking, including that concerning types, was Platonic. There were ideal types, and variation was something fuzzy. Fixity was more the rule than the exception. But after Darwin individual variation grew as an important notion in terms of development and diversity. Building off Darwin, we can see, Lycett says, that culture can evolve from variation as well, from social interactions and teaching, including "copying errors, deliberate tampering (i.e., innovation), or differing skill levels across individuals" (147). Lycett believes that the production of prehistoric stone tools is from cultural evolution or the produce of copying. Even if an individual creates material culture or art, a viewer or group is implied.

Concerning the group and individual, Nancy Aiken says in "Aesthetics and Evolution" that "cooperation and individualism" were behaviors pitted against each other, but both function as important adaptations. Through individual expression, or the appreciation and participation with such, one was able to become part of a larger group. Art evokes strong responses such as fear and pleasure, and any such basic emotions are of course universally shared. An individual, then, was able to broadcast group feelings with his or her singular expressions.

Research as old as 1924, Aiken says, has determined that fear in visual art is aroused with eyespots and zig-zag lines, especially if sharply cut as opposed to rounded lines. Visual art, then, can be used as an individual signal or as a group badge. Though a contemporary manifestation, one might want to consider Picasso's *Demoiselles d'Avignon*. Pupils dilate in fear and when exposed to sharp, pointed, or angled lines there is a similar reaction, possibly related to alarm call adaptation. We can find, however, such visual excitement pleasurable. While our sense of smell has diminished considerably compared with other species, and although our eyesight is not perfect, we are a visual species. This is to say that visually symbolic stimulation not only affects cognition but also affects individual activity and group solidarity.

The best reproductive strategy is not simply to reproduce but to reproduce offspring with longevity. For this to happen, Aiken says, cooperation is imperative, and artistic practices and learning enable such cooperative behaviors and survival. With the onset of language and cultural practices there arose a climate to promote greater understanding of others in order to cooperate. Today, we see art as a means to comprehend other cultures. Stand-alone creative behaviors, which flourished into symbolic arts, became a means for learning sociality and cooperation to re-shape selfish behaviors.

Mother/Infant and Making Special

Ellen Dissanayake is one of a few on the forefront of exploring the prehistory and ethnography of arts, so let us consider her ideas from several of her publications. In her book *Homo Aestheticus: Where Art Comes From and Why*, Dissanayake says that aesthetic appreciation is part of our biological makeup, for one can readily admit that if the arts are removed there is a mental deficiency. Just as we are hungry to guess other people's thoughts, we crave visual stimulation. In large part, there is a prerequisite for art as a behavior. Many human adaptations, such as where and how we live, are now culturally determined, which is to say are evolutionary. Most human beings prefer comfortable, safe, and decorated habitats, just as other species are instinctually careful in constructing nests and sleeping quarters.

The point is that, as other authors here discussed demonstrate, there is interplay between evolutionary biology along with its adaptations and the influence of culture. Dissanayake labels our strong requirement for culture as "bi-obehavioral" (15). Some of our species inherited traits, such as bonding, copying, caring, and learning enable us to participate in a culture. Across the globe we all bond and learn, but in different ways. In many respects this picture is Darwinian, for it reveals not the loner fighting on his own but the most adapted individuals who through variation, competition, and inheritance have survived and work together: inclusive fitness.

Dissanayake reminds us that art typically makes us feel good. Rather than a post-modern deconstructionist view that might place emphasis on the intellectual pleasure of delaying or dismantling meaning, she suggests that our aesthetic response to visual art or music is immediate and physical. From an evolutionary perspective, what we choose to provide pleasure is what has survived over millennia of our existence and is therefore an adapted behavior. Of course there are exceptions, and we can cite examples of bad behavior that makes us feel good, such as addictions that provide no benefit. Dissanayake suggests that, depending on the circumstances and context, art is a behavior.

There are innate capacities of communication and play in six-week-old babies. Dissanayake says that baby talk exhibits arrangements which are inborn since they are traits stemming from ancestral mothers who engaged offspring in stimuli that are connected with the position and movement of the body. Certainly, infants who were responsive received more and better care. Dissanayake says that our predisposed sensitivities capitalize on the culture in which we are born and help us make art, stemming from how the mother signals and expects participatory behavior from an infant.

Building off early infant-mother interactions, artists are capable of shaking the homeostasis of consciousness. An artist's work can trigger a salient emotion by attracting one's attention to shape and manipulating feelings. We are not by nature analytical thinkers but irrational. We subsequently reason about emotional stimulus and mental representations that often arise unbidden. Our mind sharing, which begins from infancy and first eye contact, is related to how we appreciate and participate with the emotional aspects of the arts.

Ceremony and ritual are forms of art and clearly, evidenced to this day, served a function to help individuals and groups survive, asserts Dissanayake. The art involved in ceremony and ritual was extraordinary and elevated one's state of mind and feeling. Dissanayake convincingly asserts, then, that art is the behavior of what she calls *making special* – i.e., "embellishing, exaggerating, patterning, juxtaposing, shaping, and transforming" that carries appeals to both emotions and cognition (53-54). Dissanayake has also referred to making special as artification, where emphasis is on performance. In this way, even art culture that is out of the ordinary, that challenges because it is weird or bizarre, fits the definition of making special. And we could see this latent in what was said of the stone tools, putting aside for a moment consideration of sexual selection.

This is why Dissanayake has coined the expression to characterize us *Homo aestheticus*. We are the species that enjoys making or participating in that which is unusual or distinctive. We just happen now to call such special creations art. We see this, too, even in extraordinary athletic performances. We find geometrical patterns and images across cultures, past and present. While geometry is revealed in nature, its lines are not as sharp, symmetrical, or straight as those we make. We find circles in nature, but ours are perfect because we make them so. There is biology at bottom here, and most likely a co-evolution of genes and culture. Generally speaking, we like patterns and order, but we also revel in that which surprises us since our brains like to be challenged and stimulated by play.

Observing the patterns and shapes in nature, says Dissanayake, and then fashioning them into something special is a means for us to exercise our need for control. All organisms in some way will strive to regulate their environments or their bodies. In our case, especially, this desire for control is mental, since from early on we were able to control our environment. We especially like to influence or control the thoughts and feelings of others, and we can do so through art culture.

In all, Dissanayake makes a very strong case for the innate human tendency to shape something exceptional from the ordinary, so that what we call art is not an extraneous by-product but a basic human function to influence or express our emotions. Steven Pinker, however, has asserted that art is a byproduct of our pleasure motivation and our technical ability to make pleasurable objects.

Pleistocene Landscape Preferences

In a 1993 survey of over one thousand Americans, Vitaly Komar and Alexander Melamid, Russian émigré artists, discovered, Dissanayake tells us, that people preferred painted alfresco scenes with mingled colors. Those surveyed also preferred to see in such paintings animals and human beings. The preferred colors were blue followed by green. Red colors can be intense or even threatening. This survey was later repeated in nine other countries spanning Asia to Africa, and there appeared a universal pattern of preferences for a natural environment containing some trees and water, similar to a savannah but not a forest.

Favored landscapes, Komar and Melamid found, seem to possess some complexity but yet coherence. As Milan Rajković and Miloš Milovanović suggest, a preferred visual composite is somewhere between the extremes of randomness and uniformity. In addition to being comprehensible, viewers liked some mysterious aspect such as a distant horizon beckoning for exploration. Likewise, Richard Coss has examined our evolved perceptual biases. We prefer canopy trees and have a predisposition for watery surfaces. In fact, we might like highly polished objects since we evolved around glistening lakes.

Stephen Davies is doubtful of the emphasis evolutionary psychologists place on the savannah hypothesis. Rather, he points out how we've adapted to many types of environments. Just as the evolutionary psychologists, Davies complains, place too much emphasis on female beauty, they do not account for how we find animals attractive. Our long history is tied up with many plant and animal species, not just men responding to women. As I suggested early on, art

adaptation most likely consists of a number of evolved behaviors, not one.

Further, Dissanayake says polling contemporary tastes in
art is of little value since art is not just an adaptive preference. One needs to explain why we have any evolved capacity for aesthetic experience, a heightening of our emotional and cognitive responses. Art is a complex response
and not merely a stimulus cue via adaptation. As Aiken says
in *The Biological Origins of Art*, in evolution what we now
call art probably had little to do with ideal beauty. Other
thinkers have commented how high European art from the
eighteenth century onward is an idealized version for an
elite audience developed from many precedents in material
culture.

Komar and Melamid present an ideal painting containing
an aggregate of what most contemporary people want in a
landscape. We also have programed responses about sex
and heroes seen in formula novels and films. We cannot say
for sure, though, that our ancestors would respond as do we
to the same painting, Dissanayake says. An art experience is
more than a reflex. Nonetheless, art does trigger shared
emotional and cognitive responses across cultures about our
biological interests, such as human relationships, birth,
death, growth, romance, family, or tragedy. We respond
positively to bright and clear colors, which suggest life and
vitality, and we respond negatively to dark and drab colors,
which suggest death and sickness. We also find compelling
any lines that meander or crisscross, since they signal danger.

But by themselves, Dissanayake warns, these types of elements alone are not art. Concerning art and aesthetic experience, the basic ingredients need shape and pattern, what
Dissanayake refers to as making special. That is, beyond
mere adaptive preferences there needs to be, as we see in
highly ritualized ceremonies or performances, some signifi-

cance about the artwork. Art is not just a display of adaptive function, like the peacock's tail, but the making of shared and collective meaning, even if generated by an individual. Art not only draws in but holds our emotional and intellectual attention, and we understand in the aesthetic experience that the artist really intends to mean something.

Art Defined

In "Art: The Replicable Unit," Kathryn Coe explores the definition of art by looking into the transition between the Middle and the Upper Paleolithic periods of our prehistory. For Coe, the implied learning and conditioning involved in making and appreciating arts served a cooperative function that helped perpetuate a lineage. What we now call art did not spring wholesale overnight. Instead, there was a long prehistory of cultural practices that helped art behaviors to flourish. Such thinking does not necessarily negate the adaptive function about art. What we define as art was in its early stages material culture, so the art making behavior of 2.5mya in stone tools slowly progressed into other media.

At first, there was body painting, emanating from a predilection for color apparent in other species. Such body painting is color and form to attract attention and therefore to affect another's behavior, which no doubt had reproductive benefits. For instance, regarding the origin of art as a social and cultural behavior, we will find that any such practice has been copied and duplicated, but the first known occurrence is most certainly not the first instance. However, people tend to imitate behaviors that are not necessarily fitness enhancing, such as tobacco smoking and over indulgence of alcohol.

Coe goes on to propose that in understanding art we should apply the definition by negation test. She believes, for instance, that symmetrical hand axes did not simply serve a functional use. Coe also looks at the human fossil record and notes that two male crania, from about 70kya, at Shanidar (Erbil, Iraq) reveal artificial head binding deformations. During the Upper Paleolithic, furthermore, we find deliberate teeth filing or removal. Later during the Upper Paleolithic there is more extravagant dental and cranial alteration, where we find precious stones and metals in teeth, which are also either incised or shaped. This behavior spans Europe, North, South, and Central America, Africa, Australia, and Asia. Coe's point is that such elaborate bodily decoration is art since its only function is to attract attention.

Concerning color, Coe reports that about seventy five pigments, from yellow to red and brown, were used by *Homo erectus* at Terra Amata (Nice, France). These colors were manufactured from various compounds, such as ochre, clay or sand, shale, and sandstone. Black manganese would also have been used. Such artful material has been found on bodies in burials, in teeth, on the remains of a deer and deliberately prepared mammoth bones at Moldova. Generally speaking, Coe goes on, the red pigment is found on bodies of mature males and also used on walls or intentionally scattered around.

Coe's main point is that the red ochre pigment develops along with bodily modification and decoration, as well as the use of beads and pendants. Taken together this is noteworthy. Against an adaptive hypothesis, art might have sprung inadvertently from other behaviors, such as the use of a cradle board or basket that unintentionally deformed a skull or tooth decoration from changes in diet and tool use. Color was already readily available from soils and rocks. Nonetheless, symbolic artifacts hold and carry ideas, values, and memories. Thus, the art-making behavior was existent even if some accidental occurrences gave rise to new manifestations.

But as other thinkers here have noted, some early instance of what we now call art could simply have been the byproduct of another process or behavior. Since the behavior was notable, implies Coe, it quickly became methodical and routine, which means that aspects of coloring, filing and shaping became technical achievements of some social meaning or significance. Costs were involved. There was pain and physical loss in any cranial or dental modification; color on skin makes one noticeable to others and therefore a target.

In support of Darwin's theory of sexual selection, Coe notes that in burials of the Middle and Upper Paleolithic and the Mesolithic, male graves contain more items and red pigments, and in later graves men have more cranial deformation or dental ornament. Darwin speaks about how through the course of inheritance, certain selected male features (color, horns) can be passed onto the female of the species. In a highly socialized culture, as that found in human beings, some traits, as Coe notes, can be passed on epigenetically.

Neanderthals and Art

The absence of widespread creativity and innovation in Neanderthals forces us, say Steven Kuhn and Mary Stiner, to consider what inspired artistic practices in later people. They believe that any creative enterprise calls on higher intelligence. While a crucial development springs from an individual, archaeologists are more interested in widespread, copied behaviors encompassing great time. The irony of sorts, Kuhn and Stiner note, is that for prehistoric practices we are therefore interested in what is copied by many but created by a handful.

We find that there is considerable steadiness over long periods of time in technological human evolution and that much later periods of fluidity demonstrate a sudden variance. This is not to say behaviors, practices, or technological innovations are fixed. Evolution depends on variation. For instance, Mousterain blade and lithic technology is quite varied and sophisticated though stone tool design is stable for hundreds of thousands of years. In many cases, it's not a result of a neural leap but simply a change according to conditions, resources, or mobility.

A good example of tools developing in concert with behavior can be seen in how food was harvested and prepared. There is, Kuhn and Stiner say, a fair amount of constancy in the diet of early people, and this stability might reflect group size and customs. There was much sharing in small groups and no need for storage. In contrast, for modern human people, food technology is in complicated tiers and networks since there are impacts on huge populations and varying cultures. There tends to be a creative, technological explosion if there is a heightened and increased social value placed on any item, e.g., from a basic sneaker to designer-athletic footwear.

In terms of Middle Pleistocene, it may simply be, then, that technology was not part of a social domain where small groups functioned in a regular homeostasis and had no pressures for tool advancement over what already worked well. Just as organisms find an environmental niche, perhaps tools do as well. I am not aware, for example, of much significant development in the fork, whether as a cooking utensil or as cutlery, in its 2.4ky history. Moreover, not all cultures use forks.

According to Thomas Junker, the oldest true art objects from AMHs date to about 36kya and are from Europe. Is there a connection between the artistic abilities of our immediate predecessors, over Neanderthals and *Homo erectus*, and our rapid flourishing? AMHs appear in Africa at 200kya and migrate out of Africa by 70kya, and art culture seems to be a significant development in human history.

But Junker makes sharp distinctions between AMHs and Neanderthals that others, such as April Nowell and the team of McBrearty and Brooks have convincingly diminished. Additionally, Junker seems to stress a Eurocentric view of art. Junker says, for example: "Art is...the only fundamentally new characteristic that the ancestors of today's humans possessed compared to earlier and other hominids..." (172). But this is not entirely true if we consider ornamentation and ritual burying done by Neanderthals.

A question arises, alluded to earlier, about the dividing line between utility and artistry. For early people material culture was functional. When did a stone tool become aesthetic? Junker sees only the more recent artifacts as aesthetic and the older tools merely as useful devices. Is beauty of functional form only modern? Probably not.

Toward a Definition of Prehistoric Art

And so we return to the familiar question: What is art? Junker gives four criteria: 1. striking shape and originality; 2. disregard for usefulness; 3. symbolic significance; 4. some degree of make-believe. Following these criteria, art need not always be aesthetically pleasing to everyone. This capacity for making art arises when AMHs split from Neanderthals 600kya but at some point before migrating out of Africa at approximately 70kya. This leads us to the questions we have been trying to address here: Why make art? Does art provide any adaptive function? Would early people have consciously made *art*? Surely something important in our cognitive development happened in those years in terms of any combination of environmental responses and brain modularity.

Junker brings into his discussion thinkers included here, such as Aiken, Dissanayake, Miller, Cosmides and Tooby, and Brian Boyd. What is the function of art if it arose and persisted because of neurobiology? Junker offers two answers: 1. what we call art had some survival value, both for individuals and especially for groups; 2. art worked as a signal in mate choice. Junker's view is that art was initially implicated in sexual selection and only later became social culture. These would have been costly signals with an advantage, and so over time they were selected for.

Darwin discusses signaling display, as we have mentioned, in terms of the propagation of elaborate colors and sounds in male birds. With human men, such signaling would have come in the form of risk taking and the ability to endure stress and strain. So, to produce an elaborate and heavy stone tool is such a signal. Likewise, men would prefer traits that are costly and rare, such as symmetrical body form or face and rich hair. Amotz Zahavi has discussed these sexual selection features as handicaps, since they indicate that if one can produce such voluptuously thick hair at such cost and energy the good health signal must be honest.

Finally, Junker sees art as an extended phenotype. This means that art can go beyond skill and demonstrate the artist's life history, moods, ideas, and emotions. Material culture and art provide crucial information about the artist's genes. Art is often shared, so it provides the important glue binding a particular culture, its traditions and rituals. In such beneficial commonality art renders the group a "superorganism," says Junker. Cooperation within groups, as well as sexual selection, provides a reasonable account for the origin and adaptive function of art culture. Furthermore, conflict between groups accounts for distinct art cultures.

Paleoanthropologist Richard Klein, too, notes that the hand axe functioned as a means of sexual selection to attract a mate's attention. Klein's thinking suggests that the man's signal would not only have been to advertise his physical and mental abilities to find and manipulate such stone. There would also be an announcement of his craftsmanship and attention to detail. These are signals about paternal competency and aesthetic vision. For some thinkers, still open is the question about aesthetics in early people. Certainly, there is, underlying the art culture itself, an aesthetic sense that from early times helped hominins manufacture shapely and balanced objects.

Klein's import, overall, is that at some point the evolution of the human brain reached a level where adaptations were to the cultural practices and artifacts themselves. Art behaviors, whether ritualistic or plastic, could be manipulated and altered more easily to affect behaviors. So while art theories depending on sexual selection are compelling, fundamental cognitive aspects are perhaps equally or more important.

Cave Painting and Superstition

Jean Clottes has written prolifically on the subject of cave paintings, but here I focus on "Why Did They Draw in Those Caves?" Drawing from Edward Tylor, Clottes suggests that cave paintings might have sprung from dreams about alternate realities and worlds, inferences depicted without a fully developed written medium. Clottes further suggests that the paintings might, in some way, address ultimate questions and posits, rather than the nomenclature *Homo sapiens*, we adopt the name *Homo spiritualis*.

Clottes asserts that such art is a symbolic representation of spiritual results, further naming us *Homo spiritualis artifex*. For prehistoric people, in an ever-challenging environment, there were constant continuities between themselves and the natural world, particularly animals. Without scientific knowledge, any explanation of the world would become more complex and varied. Their world would have been porous so that "supernatural powers" could impact them and some people could enter into this other realm, he says.

On this note, let's take a look at religion in terms of art and adaptation. Scott Atran suggests that cognitive byproducts or our ancestral need to understand natural occurrences gave rise to rituals, which in turn produced religions. We see that across cultures there are rituals and religious practices that engender, primarily cooperation in the group, and subsequently conflicts between groups.

Because we developed a theory of mind we tried to guess what others are thinking, and often this led to false beliefs. Atran says any such counter-intuitive thinking would be related to how we used what we now call superstition to explain natural occurrences. fMRI studies show theory of mind brain regions are involved in discussions of God, says Atran. Counter-intuitive beliefs, Atran goes on, are easily accepted since we are social-cultural creatures and are motivated toward our group. We accept in-group symbols and believe in-group ideas.

Atran also notes that we participate in rituals at a high degree of cost, whether in energy, time, or resources, with the single benefit of group inclusion among those who hold the same beliefs or superstitions. Adaptively, we needed to accept false beliefs to become part of a group, a function of cultural group selection. Finally, says Atran, while similar superstitions can bind, intellectual and physical competition arose between groups with dissimilar beliefs. Like religious beliefs and practices that stem from our evolved cognitive mechanisms and pro-social behaviors, so too art culture. More complex rituals developed when we moved from hunting and foraging groups to agricultural societies.

Cave Art and Moving Images

Clottes says that most Paleolithic Art was on outside, exposed stone, and has since been lost. There was a certain mindset necessary to make such art, and much time was needed to find and prepare materials. Inside caves for pre-historic people were seen as other worldly, beyond the natural realm, and risky. This suggests that inside cave paintings or engravings constituted an act to challenge, overcome, or meet the supernatural, perhaps to harness the power assumed to reside there, claims Clottes. This explanation might not be the most parsimonious. Since young men appear to have been the cave painters, the costs involved might indicate some type of mental and physical test.

The many animals depicted in cave paintings, Clottes tells us, are daunting and less hunted creatures (e.g. Chauvet). Later developments suggest that, he says, "proportions became insignificant" (e.g., Lascaux, Miaux, Altamira). At Cantabrian Spain, the doe is most typical, an animal according to Clottes associated with myth. Some perfectly good walls, in more extensively used caves such as Chauvet, were not marked, which has led to speculation that the artist needed to take advantage of the shape, texture, and contours of a panel, to feel akin to it. The walls themselves were magical, believes Clottes, evident from the application of paintings, the use of bones, and "unmistakable traces of people touching the wall..." without painting.

Art is a shift in/of attention from individual to group. In some cases the viewer is implied or imagined by the act of creation. Likewise, a viewer attempts to guess what was in the mind of the art creator. Martin Gray has written on cave art and the evolution of the human mind under the noted philosopher of science Kim Sterelny. The next few pages use factual information from Gray's paper to propel my discussion.

Gray says that cave art functioned to hold data about the natural environment externally, previously only in the brain. At the time of his research, Gray says there were 350 caves discovered, from southwest France, northeast Spain, and even in the Ural Mountains, west Russia. France and Spain have 95% of cave art. What is often left out of this discussion, according to Robert Bednarik, are the rock arts of Siberia, China, and India. Certainly, then, prehistoric art behavior was prevalent and widespread. Bednarik identifies in caves in India cupules, or petroglyph cuttings, most likely made in the Lower Paleolithic with Oldowan tools, which therefore dates art making as very early and not confined to Europe.

Portable art, such as engraved bone carvings, can indicate a nearby cave. The Upper Paleolithic or Late Stone Age runs from 40kya-10kya. AMHs such as *Homo sapiens* were in the west of Europe by circa 40kya, while Neanderthals had been there already for about 200ky. This means, for the most part, that any art culture is probably produced by modern human beings. Genevieve von Petzinger and April Nowell, however, see a problem with dating the art in Chauvet, which might not be an anomaly. They place this cave art in a tradition that began in Africa and the Middle East placing symbolic expressive culture to at least 100kya. Principally, von Petzinger focuses on the dots, line patterns, and cruciform figures in Chauvet. These graphics are geometrical, non-figurative symbolism of an older practice.

Of course, art persists even after Neanderthals. Paintings at Chauvet cave date to approximately 36kya. Dating is easy for organic material on, near, above, below an object, such as a stone tool. Pollen deposits are also used for dating, but these are problematic with exposed cave paintings. We can examine the subjects of the paintings themselves, such as the wooly mammoth and other extinct animals to help in dating. If a cave painting was sketched in charcoal, as some are, radio carbon tests can be done, as well as testing fibers that might have been applied along with the paint. Often, carbon deposits from torches used for cave lights can be tested with radio carbon. Recent tests reveal that some of the pigments contain ochre and a coagulant like urine or animal blood.

Ochre has a long history in human culture, going back to circa 900kya in South Africa. This compound was used for numerous purposes, ranging from the medicinal, as a binding agent, and for bodily ornament. Ochre in India from at least 300kya appears to have been rubbed onto rocks apparent from scratches on the ochre. Ochre has been found in the Czech Republic circa 300kya, in an area where it is not naturally occurring, so it was sought and carried there. Other painting compounds used include manganese and iron oxides, black and red, the colors of most wall art, though other colors were available. There were costs of time and energy to find and make paints. A logical question then: What was the benefit?

Gray says that torches used as a light source in the caves were also rubbed onto walls. The light from the torches did not last long, so lamps fueled with animal fat or tallow and a crude grass wick were used. About one hundred such lamps have been found in Lascaux. Here too there were costs involved in utilizing such light energy, although dim. Why expend such effort to paint inside a cave? Some painting did occur near the opening of caves where natural light was available and the finished product would be accessible to many viewers. Other paintings within caves are often difficult to access, possibly only for artist and a select few.

Some artists employed a dappling or stippling effect to make use of contours and other aspects of rock walls, which were often scraped in preparation for art, though in some cases paintings were done on top of a previous depiction. Generally speaking, there were no major style shifts over the course of time. Most paintings are of animals, herbivores, horses, but the oldest (so far) cave at Chauvet has predatory beasts, lions and bears. Importantly, the animals depicted in the paintings were not necessarily those of the area or those hunted and eaten.

Evolutionary psychology tells us that in terms of mating habits, young men take risks to display their fitness. The danger of the caves might have been a means for such male advertisement. Handprints from the caves have been analyzed and the conclusion is that the art there was produced by adolescent boys. Art in remote caves could then be part of so-called young male syndrome, evident even today in risky behavior to out-compete rivals.

But for handprints, there are no real human depictions in cave art, and what has been found is disputed. There are handprints from male adolescents about ten to sixteen years old. Other markings also appear as either early writing, symbols or tally marks, circa 27kya, at Cuevadela Pileta, Spain. Paintings, engravings, and even some clay sculptures are amazingly detailed and accurate. Several form panels. Some speculate that the best artists were honored with cave space. Cave bears, which were very large and dangerous, became extinct by about 28kya and perhaps why no cave art appears before 32kya. Considering the perils of entering a bear's lair, whether present or in absentia, might also explain why art is there.

Because of dramatic climate changes, such as glaciers that would have destroyed caves, as well as exposure to light, animals, moisture, bat droppings, pollen-infused growth, we probably have only a small portion of all Paleolithic cave art. The inside of the caves preserved what was undoubtedly lost on the outside. Likewise, other artifacts from antler, bone, and stone, thousands of such figurines, survived because of the durable material used. Wood and clay artifacts no doubt succumbed to the elements. To emphasize, artistic behavior in prehistoric people was not uncommon.

Here are some examples. The Venus figurine of Hohle Fels cave in southwest Germany dates to approximately 35-

40kya (early Aurignacian). Stephen Davies says such fat Venus statues are less what Pleistocene men preferred and simply represent a pregnant woman. Demonstrating our predecessors' long history of closeness with animals there is a horse figure fashioned from ivory at the cave of Vogelherd, Germany, which dates to about 33kya. A lion's head on a man's body of ivory at Stadel cave, Hohlenstein, Germany, dates from approximately 30-34kya. In addition to its prevalence, early art culture demonstrates a high degree of crafting skill.

The most and oldest cave art appears only in Western Europe, but this might arise from two facts: 1. There are more caves across that part of Europe. 2. Since it is a more densely populated area the likelihood of discovery is greater. According to McBrearty and Brooks, AMHs who were, much earlier, in Africa also demonstrated cognitive abilities before those seen in Western Europe, such as blade technology, bone tools, and trading practices. Neanderthals, for example, buried their dead, used pigments, made jewelry, used tools, and had charcoal, but there are no paintings from them as far as we can so far tell.

Art and the Human Psyche

So, we come back to our original question. Is art an adaptation? To help us answer, let's turn to biology. There are three changes in organic evolution. 1. Adaptation for fitness, survival, or reproduction, and such adaptations are heritable. 2. Byproduct (a non-adaptive side effect) not related to fitness. 3. One-time mutation that could possibly but not necessarily be related to fitness. An example of a byproduct would be that blood is red; it need not be. Or that calcium makes bones white; color makes no difference for the strength of bone. And of course some say that art is a byproduct of other fitness-enhancing mechanisms, such as mind or learning, and not an adaptation itself.

The origin of art might be psychological and not necessarily directly related to fitness, survival, or reproduction. This is to say that art culture arose to satisfy a mental need to overcome a bad feeling or to explain something unknown. There is also genetic drift, which means that some individuals by accident have more surviving offspring than others. There can be bias within a population because of, e.g., a random environmental mishap and not simply because of fitness, though this typically affects small populations. We could apply this line of thinking to cultural productions. Even if not fitness enhancing, someone or some group could produce a novel behavior that is admired and therefore copied and imitated. Clearly, though, the evidence so far suggests art behavior as an adaptive mechanism, a non-random variation.

Gray reports that David Lewis-Williams, and then supported by Jean Clottes in 2008, suggests cave art was part of some hallucinatory ritual (see, e.g., *The Mind in the Cave*, 2002, Lewis-Williams). This theory does not appear as widely accepted as those dealing with social or sexual selection or cognitive enhancement. Evidence suggests that caves, some quite small, were not used by many people, so what is the ritual? Much of cave art is in small spaces, and so any hand prints found could have been private display and not for religious reasons. Such high-cost cave art might have had something to do with adolescent maturity, sex, and risky hunting games.

Many early finders of cave art, incidentally, were in Roman Catholic countries, so their interpretation of what the visuals represent is biased. Early explanations, says Margaret Conkey, attributed the paintings to magic since such evaluations were based off Eurocentric art history. The majority of the art is probably symbolic material culture and not with any particular ritualistic or religious significance. Prehistoric, hunter-gatherer art is material culture, meaning that function and symbolism are entwined. The depiction of the animal *is* the animal.

Beauty and the Brain

There are two main periods of human brain increase: 2-1.5mya and then 500-200kya. At these junctures there are peaks in cubic size as well as a combination of tool manufacture, tool use, and social grouping. Size alone does not imply higher intelligence, since the Neanderthal brain was as large as that of an AMH. What counts is modular flexibility, neural connectivity, and networks. Some essential cognitive competencies would include a number of functions. 1. Attention, both individual and shared. 2. Awareness of the environment and of others. 3. Self-awareness. 4. Memory. 5. Symbolic thinking or internal representation that carries meaning and can be shared. Cave art could be a byproduct of a combination of these cognitive abilities as a prompt to take representations from one's mind to be communicated to others, so there is some type of social component to art.

Tooby and Cosmides ask the question, in the title of one of their papers, "Does Beauty Build Adapted Minds?" In all cultures we find a need for imaginative engagement. The creation of artistic works is not an evolutionary blip, for human populations over tens of thousands of years have found such creations pleasurable to observe and rewarding to make. Such involvement with material art culture on both the creative and participatory levels could be possible only when we consider that the mind is adapted for such.

Consider how often many people respond positively to photos or depictions of sunsets. E.O. Wilson has gone as far as saying, as posited in his biophilia hypothesis, that human beings have an innate attraction to the natural world. Research by Omid Kardan et al. demonstrates our need for and deep connection to green spaces. The nature of the evolved mind is creatively cognitive and representational. We'd never have been able to reach our level of sophistication had it not been, and so there are some cultural universals. Likewise, play is inborn and essential, permits the child to test limits of self and others. So, too, art is play in that it lets us test limits of perception, thought, and inventiveness for social display.

The evolved human mind has in fact been adapted to consider imagined worlds, or what Samuel Taylor Coleridge referred to as the willing suspension of disbelief. For Tooby and Cosmides we are aesthetically pleased or aroused by images since they cue to our survival and reproduction in the long Pleistocene environment in which we evolved. What we call beauty might have been, evolutionarily, a mental tuning mechanism, they say. In this way beauty is both a quality that invites attention as well as something perceived as having value. Of course, there can be individual and cultural differences here, especially in any value laden appraisal. Some groups go as far as trying to eradicate a competing group's cultural icons.

At any rate, say Tooby and Cosmides, representations are stored across brain structures associated with experiences and emotions. Reading neuroscientists such as Antonio Damasio and Michael Gazzaniga will also reveal how emotions are primary and tied to decisions and memories. Since we evolved from earlier species that had to deal with predators to family and self, much of what we hold in terms of value is emotionally dense. Tooby and Cosmides go as far as saying that art is a universal since we have all evolved to make art. But again, we must be careful how we use the word art. In the past few centuries art equates to high art for the elite, whereas art making behavior for prehistoric people served a functional, material purpose. This does not eradicate, however, an aesthetic instinct in our species.

The Sound of Art

Let's shift gears somewhat from visual art to music. Steven Mithen, in "The Music Instinct," says that a combination of artificially fabricated and naturally occurring vocal sounds that could be varied in tone and tenor were a necessary component of communication before language. Consideration of music is important, in part, since like visual art it is material culture with a social function.

Because of this pre-linguistic musicality, we now have an instinct to make and listen to music. Every culture, past and present, has some type of music, as hard as music is to define. Almost everyone everywhere has a strong and lasting connection to musical sounds and rhythm, says Mithen. We evolved musicality as essential to survival and which in turn helped evolve the cognitive functions of our brain.

Mithen says that music for us is biological and not just cultural, apparent, for example in the musical sounds made by babies even in utero. We have brain regions, such as Brodmann's, solely for processing music. Apes and monkeys, who are in our evolutionary lineage, express themselves musically in their many and varied vocalizations. Chimpanzees are known for their drumming displays. With *Australopithecus afarensis*, one of our early ancestors from circa 3.5mya, there is reduced teeth size, which equates to a larger oral cavity and therefore the ability to make more varied vocal sounds. The facial difference in chimpanzees renders them incapable of enunciating some of our vowel sounds.

Key to our discussion is how these vocalizations, even among other primates, are crucial in emotional expression. Mithen's point seems to be that emotions, which helped make survival and reproductive decisions, were as they still are vocal and expressed musically. There is a brain/body sound sensitivity similar to visual sensitivity. As with the reduction in tooth size, bipedalism changed our vocal anatomy since the re-direction of the spine lowered the larynx, which lengthened our vocal tract and so an ability to create more sounds. Bipedalism also made it possible for us to control our breathing better in terms of singing, and all of these changes were already in place with *Homo ergaster* at 1.5mya.

Migrations out of Africa about 2mya for the first wave of *Homo erectus* enabled them to encounter, in Southeast Asia and Europe, many new bird and animal sounds. Even today many languages mimic animal sounds. As groups enlarged and demand for cooperation increased, with big game hunting at about 500kya, Mithen postulates there was much "singing and dancing together" as a bonding mechanism, derived from *Homo ergaster*. Neanderthals by about 350kya, with whom we likely share a common African ancestor circa 500kya, *Homo heidelbergensis*, probably sang, says Mithen, like the more primitive ancestors.

Neanderthals left no substantial artistic culture, no paintings or other such art, probably had little sophisticated spoken language, and used the same type of stone tools with negligeable innovation for tens of thousands of years in spite of adaptive and environmental pressures. Nonetheless, Neanderthals did have material culture, used ornament and decoration such as beads, and buried their dead.

However, Mithen says Neanderthals surely communicated as did all of our hominin ancestors and most likely evolved what he calls a "proto-musical" means of communication, Hmmmmm. According to Mithen's plan, these sounds would be *h*olistic phrases, *m*anipulative in that the goal was to affect another's behavior and not transfer information, *m*ulti-*m*odal since the body and the voice were employed, *m*usical with tones of emotions for a variety of caring or sexual displays, and *m*imetic in how sounds copied what was heard in nature.

Mithen says the same system of Hmmmmm was used by our African AMH ancestor of several hundred thousand years ago but then split into both language and music. By the time of the second out-of-Africa migration, circa 70kya, the differentiation between music and language helped us evolve and survive in ways the Neanderthals could not.

For our purposes, musicality, which we now see as art behavior, was originally adaptive and served several key functions. That Mithen sees musical vocalization as powerful enough to influence another's behavior is akin to how visual arts, too, can cue emotional and cognitive effects stimulating behavioral change, even if temporary.

Like music, visual art had some connection with the natural world, and not only in what was represented. Consider how cave painters synchronized natural materials with the texture and curves of the walls, and how rock art depends upon surface dimensions. Obviously, our species had a very deep aesthetic impulse derived from nature, and what we see in prehistoric cultures is not the birth but the flourishing of art behavior. Art making precedents and capacities thus run deep in our prehistory and must have served any number of adaptive functions, from mental, social, and sexual to persist and thrive.

Chapter Three

OBJECTIONS

W. Seeley, both a philosopher of art and a sculptor, says that skeptical objections to any scientific study of art point to how an aesthetic experience is not reductive. Seeley disagrees with such skeptics and says neuroscience merely explains the effects of perception in our visual field. There is natural beauty all around us, so what, then, is the difference between a striking sunset, an inspiring star-filled night sky, and what we call art? The difference is that in human art behavior we discriminate among evaluations and ascribe meaning to work from an artist. Natural energies do not intend to be symbolic, but an artist could.

Our extended gaze on objects demonstrates the interest they hold for us, whether out of curiosity or aestheticism, but once such an object is rendered into a sculpture or painting we begin to reflect back-and-forth between the artist's intention and our own interpretation and feelings. Does this exchange carry an adaptive aspect? Some say, strictly speaking, no. Perhaps art is a byproduct of the adapted mind and not itself an adaptation. Cognitive functions, however, are very much part and parcel of art behavior for maker and participant, and symbolic forms are such potent graphic communication, it's difficult to dismiss them as mere byproducts.

Stephen Davies, in *The Artful Species*, does not wholly ascribe to art as an adaptation, and he is reluctant to buy into aesthetic sensibilities in human beings as continuous from other species. However, Davies does not disallow art culture to be functional.

What, then, is the evolutionary function of art behavior? What does it mean for a trait to be adaptive?

By 1966 a more complete understanding of Darwinian evolution was taking place. Biologist George Williams stresses that we should not haphazardly label certain traits or characteristics as adaptations. Williams emphasizes that an adaptive function must arise through the process of evolution by means of natural selection and not be a byproduct that we simply label as an adaptation.

David Buss summarizes Williams' criteria for an adaptation as the *efficiency* to solve a problem, *economy* in solving the problem, *precision* in accomplishing the target of an adaptive problem, and *reliability* in the operation of the adaptation. Readers can decide for themselves, but we are thinking of art behavior from two perspectives in addressing these concerns. 1. Inclusive fitness, where the kin group and social aspects of culture play a part. 2. Sexual selection, where the mental faculties operate to make choices.

An adaptive problem ultimately deals with reproduction since the function of organisms is to pass on genes. We can therefore think of our prehistoric ancestors developing and using artistic behaviors to address individual and social problems regarding mating strategies. There have been, nonetheless, some strenuous objections to the role of sexual selection in art behavior.

An Art Instinct?

Denis Dutton, in *The Art Instinct*, has famously asserted that any creation of art is instinctive. More particularly Dutton relies on the ideas of sexual selection originally formulated by Darwin and more recently articulated by the evolutionary psychologist Geoffrey Miller. Dutton thinks it's a mistake to separate adaptations and byproducts with human beings, since our evolution is quite complex. Art clearly has functionality in how it makes us feel good, happy, whole-some, reflective, and even sad. In other words, our needs may have given rise to the arts. Needs, desires, motives and art behaviors, whether in the making or viewing, are so closely intertwined it's difficult to separate them.

At the same time, Dutton stridently opposes any views that see art as social. Such a stance is more or less against prevailing wisdom seen to some extent in writers like Dis-sanayake and Coe who emphasize how children learn art-like behaviors from their mothers so as to foster sociality. The fact is that art *is* social. Otherwise, we'd not be talking about its cultural dimensions right now.

Dutton also opposes any view that sees material culture as part of art making. This is principally because Dutton em-phasizes high art from the Western tradition. As we've not-ed, cave art did not spring suddenly, but rose slowly from prior art making behaviors visible in prehistoric material culture. Likewise, Stephen Davies stresses how Ellen Dis-sanayake objects to Western high art from the eighteenth century onward as self-serving for an elite group.

Strikingly presumptuous is Dutton's claim that the arts are not religious, moral, or political. Of course he'd say this with his insistence on sexual selection as the driving force of art culture. While prehistorically in its earliest forms art making was likely not political, it is hard to imagine any creator who is not aware of an implied viewer. Because of its representational and symbolic nature, art culture is meant to be seen. And any creator would comprehend that not all viewers would see a work in the same light, which means that there would be disagreement amid various opinions. Art making is cultural, and culture by definition not only joins but also separates individuals and groups.

Against Dutton and art as sexual selection, not only do women make art but art is made by all age groups. We should also consider by analogy male bowerbirds who construct elaborately designed and decorated bowers to court a potential female mate. Is bower art behavior absent social significance? Probably. Even if the bower is for a few select viewers, does it then compare to cave paintings? No. Often the objective of the male bowerbird is not achieved, although his activity is indubitably driven by sexual selection.

While I am a strong advocate concerning our continuities with the animal world, especially with apes, we are certainly a different (not exceptional) species with complex motives and incentives. Moreover, in most species male bowerbirds tend to build new bowers each year, mainly because of habitat changes. For those who abandon their bowers, the implication is that there is no cognitive or emotional connection to the construction. But we know birds are sentient beings. With the bowerbird, though, any so-called display behavior clearly involves some elemental aesthetic sensibility in both the male and the female.

In the case of human beings, we do not employ art works simply to attract mates but, even as Dutton would agree, to possess an aesthetically pleasing artifact over generations. Participation is implied in art behavior. Human artwork is not, therefore, completely derivative of sexual selection. Stephen Davies objects to the arts as sexual selection. He suggests that our hominin ancestors might simply have made aesthetic choices based on what attracted them as beautiful and calming without any thought concerning the selfish gene or extended phenotype.

As I've been saying, human art culture involves a constellation of behaviors ranging from the sexual, social, and cognitive.

The Ancestress Hypothesis

Kathryn Coe offers an intriguing objection to any emphasis on sexual selection in her book *The Ancestress Hypothe-*

sis. Coe opposes any notion of art as male-based and stemming from competition. Instead, like Ellen Dissanayake and Richard Alexander, Coe sees what we call art as a scenario maker to help solve social problems by fostering cooperation. Coe looks especially at mother-infant bonds. A mother not only attends to the physical needs of her child but tends to ornament, decorate, or dress her child in special ways. Consider the multiple mother-infant reciprocal interactions. She plays with, signs to, and sings to the child. Material ornamentation could be as simple as a distinctive hair style or as elaborate as dental modification. Coe finds art behavior originating in the woman's need to identify ancestors, as a way to mark a lineage.

Rather than seeing art behavior as Dutton and Miller in sexual behavior, Coe sees art behavior arising from long-term maternal care to help the child survive and succeed socially. Key here is that such attention to how the child looks subsequently gets handed down and spreads over generations as a tradition. In this way the ancestress methodology is social and not self-interested. Kin and descendants artified in a special way can identify and so cooperate with each other.

Humanology

K.P. Mohanan coins the term *humanology* in an attempt to explain current trends. In the past, we had simply humanities or conceptual inquiry with attention to analyses of texts. Now we have humanology, which extends humanistic discourse by including the sciences, much as we are doing here. Mohanan says he would have preferred Dutton to specify which "strand of art" is a biological adaptation.

But that misses the point. Just as there is no art gene, there are rather underlying genes for learning, and so there are underlying genes for making art and music. Art is a complex of behaviors, much like morality. No single type of art is an adaptation. Mohanan sees Dutton's effort as a failure since he does not separate the threads from the entire fabric of artistic culture. But that would be looking for only one behavior to explain, e.g., a fear response. Fear is but one reaction to unexpected surprise.

While I agree that Dutton's extraordinary emphasis only on sexual selection excludes other key factors in the adaptive function of art, any total rejection of sexual selection in art making is equally misguided.

At the same time Mohanan is right to suggest Dutton's weakness stems from placing all art behavior only on sexual selection. Any instincts we have are modulated by our neocortex, which is geared to caring, helping, and sociality.

Mohanan questions the emphasis on sexual selection since the human race clearly has many examples of people who sacrifice any possibility of mating, to produce biologically similar offspring, such as those committed to a single life because of professions (nineteenth-century dons), beliefs (Roman Catholic priests, brothers, and nuns), or biology (gays and lesbians). Of course, gays and lesbians *can* and do have biological children. Priests have surrogate families. These are exceptions based on other forms of adapted behavior, such as, with priests and nuns, conforming to a group.

Social Selection Over Sexual Selection?

Brian Boyd, noted for his work in evolutionary narratology, also takes exception with Dutton. He says, for example, that original body painting was for social, status, or group identification and only later for sexual selection. Like Mohanan the objection is mainly to art only as sexual selection. Boyd, instead, sees art as a means for us to engage in necessary cognitive exercise in terms of, e.g., focusing attention or comprehending patterns. As we saw in our previous section, there is no reason why natural selection (cognitive functions) and sexual selection (display features) could not *both* be at work in terms of the adaptive nature of art.

It's impossible to claim that one gene is solely responsible for any behavior, and so too with art making. The so-called language gene, FOXP2, for instance, appears not only in Neanderthals but in other species, yet we express the gene with advanced writing and complex languages.

Ellen Dissanayake, to whom we have already been introduced, also raises objections to Dutton and his emphasis on sexual selection. In spite of Darwin's enormous contributions about insect, mammal, and bird charms, displays, colors, and songs, Dissanayake is doubtful that sexual selection explains art. Any characteristic selected for display, Dissanayake says, is at great survival cost for whatever reproductive benefit. However, some biologists might say reproduction is more important than survival. One could survive and not reproduce. Reproduction is survival of one's genes or a version of one's self.

In addition to Boyd and Dissanayake, other opponents to Geoffrey Miller (*The Mating Mind*) and sexual selection of art include (referenced to positively in other parts of this primer), Nancy Aiken, Kathryn Coe, and Joseph Carroll. Dissanayake takes aim at Dutton through Miller, saying that Dutton often mimics Miller's claims. For example, Miller says that an average adult has about 60,000 words of vocabulary that can be explained only by sexual selection, or the need to impress a potential mate intellectually. Dutton says that, in terms of this large vocabulary, we have many words for colors, but Boyd counters and says that other cultures have far fewer words for color, and at least some have only the equivalent of dark/light, which is not explicable via sexual selection.

Rather than sexual selection, Boyd says that art is a social problem-solving mechanism and that sexual selection simply magnifies what we call art. In other words, Boyd focuses on sociality as key to the arrival of artistic culture, not sexual selection, which attends only to reproduction. Boyd says the costly signal is not just sexual selection but social, i.e., how one increases his status in the group. Perhaps, but an increase in status means more mating possibilities, evident in chimpanzee groups. Sexual selection, says Boyd, focuses only on the costly. Indeed. Joseph Carroll would suggest, too, that just because an adaptive feature is attractive does not make it sexual.

Dissanayake sees artistic culture as ceremonial display that makes us feel good and connects us emotionally to a group. As Boyd and Carroll seem to think Dutton minimizes the social functions of art, Joanna Klara Teske believes that Dutton minimizes any cognitive function, which would permit explanation of the intricacies of art. I will outline such cognitive benefits in the next chapter, but suffice it to say that lines, colors, shapes, and patterns underlie our visual apparatus. In order to survive, we had to respond to many different visual cues simultaneously in our environment of evolutionary adaptedness.

The Biology of Art as Speculative?

A strong objection to evolutionary readings of art comes from Ronald De Sousa who insists that any biological explanation of art is purely speculative. Of course, this is not so. We have artifactual materials, referred to in the previous section, which anthropologists can connect to behaviors seen in contemporary hunters and gatherers. Material and then art cultures serve social, sexual, and neurobiological needs.

OBJECTIONS

Objection to any connection between art and evolution is self-blindness. For example, De Sousa mentions Darwin only once in his paper and does not cite or explain *The Descent of Man* which is the pivotal work on sexual selection and behavioral display. De Sousa does not seem, in fact, to understand adaptation and natural selection, for he applies to these processes the word *purpose*. There is no purpose in nature, only the abilities to survive and reproduce in terms of fitness costs and benefits, both individually and in a group.

De Sousa first says art is an exaptation, which is a feature that evolved for one function but then is molded by other selection pressures for a different function. See, for example how the middle ear bone derives from the jaw of our ancestral reptilian predecessor. De Sousa then asserts that art is a spandrel, something valued but undergoing no selection pressure. Whereas exaptation is an opportunistic change, a spandrel is a side effect. He suggests that our valuing of artistic culture is "pure chance."

If this were so, natural selection would have eliminated our need for arts long ago. Chance is not the answer to the questions we've raised about how and why art behavior has persisted for so long across so many cultures. Chance does not account for the costs we put into making art, the emotional and social impact of art, or the mental play and physical pleasures of artistic culture.

While mutation and some irregular effects occur in nature and factor into evolution, Darwin's emphasis is on a flowering of diversity via variation. I don't think the first edition of *Origin of Species* uses the word random, but he spends considerable time discussing variation, which can occur randomly. Mainly, evolutionary forces sculpt new mechanisms and better functions from what is already present. We can see variation in our cultural artifacts, from the first stone tools to more complex artistic culture much later. Adaptations evolve, and art behavior, whose earliest origins have been lost but not its many subsequent incarnations, fits this paradigm.

De Sousa does not believe the universality of art necessarily correlates to any function, although artistic tendencies occur spontaneously in children all over the world, and adults respond positively to patterns and forms of artistic culture. De Sousa disagrees with Frederick Turner who, quite rightly in line with Darwin, sees our aesthetic sensibility itself as an adaptation.

Putting aside the other objections, I think arguments such as De Sousa's are raised simply because what we value and hold dear is uncomfortably put in an evolutionary context. Some people object to any claim that we, our minds, or our material culture and art behaviors are the products of natural selection, sexual selection, or other evolutionary forces.

Many objections seem to rise from an unfounded fear that we are animals (we are), that we have no free will, and that all of our behaviors are determined. As a salve, might I offer to these objectors the expression *evolutionary energy* rather than evolutionary force?

More cogently, Johan De Smedt and Helen De Cruz have argued that art behavior is neither a byproduct nor an adaptation. Rather, they suggest that cultural group selection through altruistic cues and ethnic markers are responsible for art behavior. What we now call art was a method of group cultural recognition.

While Dutton and others focus on objects of art, they neglect cognitive functions, to which we now turn. The mind is a modular network. If it were not, any change in the brain would negatively affect the entire organism. Instead, modularity permits certain brain regions and capacities to evolve adaptations separately and differently.

Chapter Four

NEUROBIOLOGY and COGNITION

Mariagrazia Portera, writing about evolutionary aesthetics, says that some in the field, such as Eckart Voland, Cosmides and Tooby, and Dutton are not very Darwinian since their focus is on human only and not animal evolution, as done by Darwin in *The Descent of Man*. As Darwin suggests, and referred to in our first section, any human sensibility for what we call beauty is apparent in other species and not germane to us alone.

We see male bowerbirds who manufacture elaborate constructions, colorful birds of paradise, peacocks with fan tails, and butterflies, for instance, whose colors do not seem to operate in terms of natural selection. Such creative behaviors and colorful decorations are evolved adaptations, not accidents. Portera's idea is to find an aesthetic attitude in perception undergirding attention, emotions, physical costs, and sensual discrimination. In this way the aesthetic attitude is autotelic, or an activity that has an internal drive which is mentally self-rewarding.

There are many ideas about the cognitive functions attached to the development of animal color, lines, display, and our own artistic culture. To start, let's examine some of the writers found in Mark Turner's book, *The Artful Mind*.

Puzzling Creative Cognition

In "Art and Cognitive Evolution," Merlin Donald asserts that art is principally directed toward a mental outcome. What Donald means is that artistic culture is not necessarily the physical result of a utilitarian product. Rather, the function of what we call art is cognitive. The modern human mind contains all of its previous incarnations, since evolution can only add onto what came before. Donald says some of the ever-increasing adaptations spanning 2mya to 150kya would include the mimetic, gestures, dance, ritual, the visual, myths, and finally language.

Donald uses a forceful tone by saying that artistic culture "attacks the mind" via our sensations and emotions. Anyone or any group viewing art attempts to reassemble the artist's experience as her own. Artistic culture, which our ape cousins do not have, is a networked mind-sharing experience. This occurs since our species has a thoughtful tendency, evident in children, where we reconsider our own ideas and actions. Worth noting, however, would be work by, to name one, Michael Tomasello who sees apes as thinking creatures, as they and other species are. At any rate, in human beings a reflective inclination permits us to play with, rehearse, and evaluate our own behaviors in layers. In some ways, artistic culture engages our minds by toggling reality, thinking, and planning in a similar fashion.

For Terrence Deacon, to be human means non-literal expression. Artistic culture is not automatic, like breathing, but requires language, cognition, and cultural experience. To emphasize, culture is not de novo but evolved from us. Specific arts have to be seen as the fabric from our evolved tendency to make symbols. Even today our societies are rife with symbols, from national flags, religious icons, to corporate logos. Each of these symbols represents values and ideas, and people have very strong reactions for and against the principles such symbols represent.

Symbolic thought is a function of higher intelligence. While many primatologists, from Jane Goodall and Frans de Waal to Dorothy Cheney and Richard Seyfarth, have demonstrated the high cognitive functions of apes and monkeys, they do not seem to have any symbolic culture. This is not to say they do not have a latent capacity for symbolism because they are able to recognize and to differentiate signs and signals.

Deacon sees some mechanisms of artistic creativity not selected for but evoked from other evolved capacities, such as cognition. Perhaps. But that line of reasoning does not discount art behavior itself as an adaptation. Material art culture has solved problems, via the discussion above, related to both mating and social functions. This has evolved for millions of years, and these behaviors are inherited. So we continue to hover around the question of art behavior as an adaptation.

Although we have previously noted how making art comes naturally to children, many forms of artistic culture involve experience to develop fully. But that implies how crucial the behavior is for any individual or group, considering the cost/benefit equation. There is a human aesthetic faculty, says Deacon, far beyond the scope of what Darwin discusses in animal sexual selection, where we tend to make an object or an action as a symbol.

This is true, but to be fair to Darwin, he was a Victorian in a religious culture and therefore hesitated until the 1870s to reveal his few ideas concerning human evolution. Many of his scientific friends objected to what they perceived as his anti-religious ideas. For instance, in spite of its title, *The Descent of Man* weighs in at almost 700 pages (Penguin edition), but only 68 pages of Part III deal with humankind. Of course there are references to human beings throughout the book, but the point is that if he truly had free reign, what would Darwin have completely revealed?

Deacon, nevertheless, correctly ties the aesthetic faculty to something emotional and says the two were somehow linked in our evolution. Indeed, as neuroscientists would say, we know that cognition and emotions are closely connected. Private feeling and public emotion precede cognition, and one need only consider, for example, fear responses later regulated by thought. We share and advertise emotions to convey thoughts and feelings. These cognitive adaptations are closely allied, then, to the function of art making.

At any rate, the ability to create and experience aesthetically is somehow a broad extension of our cognitive abilities, some would say. I'd suggest that creation is simply a magnification of what is already there. Or creation is a different means of expressing what is latent. Does that negate any adaptive function? No. Jealousy is adaptive, but we have different ways of channeling that feeling. Deacon believes that a nascent symbolic mechanism 2.4mya in Australopithecus created a shift toward cognition, which led to "manipulating mental representations" associated with emotional response.

Art behavior and material culture are the physical manifestations of manipulating mental representations. Being once removed from cognition does not necessarily disqualify any adaptive function. I don't mean to press the issue too much, and I will let readers decide. The cognitive approaches to art behavior are compelling, but sometimes they address the how aspect more than the why, as we shall see.

Francis Steen, also emphasizing the cognitive aspects of artistic culture, says that an aesthetic experience itself is adaptive since it helps one in self-construction. Steen, in fact, goes as far as saying that in terms of making a self-identity what we consider beautiful is a "resource." Aesthetic experience is an evolved cognitive function since it helps to build and to maintain mental patterns and order them as attention, consciousness, and response. Certainly, we discriminate among stimuli and hold onto those we want to associate with, either as individuals or as a group.

On a related note in terms of art, aesthetics, and cognition, Mark Turner speaks about conceptual integration. Turner says that mental blending is very old and most likely arose in mammals. Much of this approach deals with understanding the barrage of perceptual material thrown at us any given moment, so we evolved the ability to compress such inputs to manage them, and this has given our species, perhaps in conjunction with sexual selection and social selection, the ability to make art. Not to privilege *Homo sapiens*.

This is to say that our brain neurobiology has evolved to seek cohesiveness. We want to be in control of our bodies and of our environment. Even with abstractions such as the Cubism of Picasso, both the artist and the viewer are engaged in a process of mentally negotiating irregularities. But this is not to say we don't like a mental contest and that we don't like to be visually surprised.

Art, Ambiguity, and Making Meaning

At the same time, Shirley Brice Heath writes about how any such irregularities arouse our curiosity. Artistic culture can be an enigma of incompletion that cognitively challenges us to attempt completion. We like to solve puzzles. Heath raises the notion that artistic culture is a form of mental play we actively participate in not only for enjoyment but to satisfy an unconscious need for brain health. One of the important components of mentalizing in artistic culture for both artist and viewer is how we are tested with un-puzzling as-if and what-if scenarios. While no parietal art has yet been found to precede Chauvet, in addition to the many theories about cave art, mental scene recreation might be one.

Cognitive play in art is a welcome form of detection where we attempt to connect the dots and literally figure out what we are seeing. As mentioned, even in cave paintings there are arrays of dots and lines. Surely any such what-if visual configurations are adaptive. This mental play, says Heath, will strive to make what is temporary permanent and what is piecemeal whole. To achieve these mental feats implies that within the mind we have the ability for completion of the as-if and what-if scenarios.

But again, as with other objections, some might say the true adaptation is cognition, whereas the art play is a byproduct. If this is so, why is the byproduct so prevalent, even in prehistoric time post 70kya, that it has a life of its own? By definition, a byproduct is unintended. In answer to all of those who see art behavior as a byproduct, how could there be so many unintended artifactual forms of material and art culture?

Is imagination a byproduct? Are mental representations byproducts? As Samuel Moulton and Stephen Kosslyn correctly ask, the key question is not what is imagery but what is the psychological function of imagery. The answer is that mental images allow us to simulate outcomes and so make assessments and predictions. Imagery enables us to take external reality into our minds for storage and manipulation. In fact, part of our evolved architecture, it seems, is precisely that we calculate levels of danger or the fruits of resources based on sketchy perceptions.

Granting that art making *might* have been, at some point very early on, a byproduct of cognition, clearly it rapidly began to have its own important adaptive cognitive functions, whether in terms of sexual selection or social selection.

Heath further says that part of our evolved understanding of perceptions seeks meaning to decide on movements. We like graphic art since it echoes the visual trials we undoubtedly encountered in our environment of evolutionary adaptedness: finding patterns, comprehending lines and color, understanding movement. As an example, Heath offers Diego Rodriguez Velázquez *Las Meninas* (1656), which shows a regal couple presented through a mirrored image. The viewer is invited to re-frame what she sees in order to complete the painting. Part of our evolved neurobiology is an aptitude not only to predict outcomes based on visual perception but also to revise our thinking about such perceptions with new experience.

George Lakoff, in "The Neuroscience of Form in Art," says that any theory about form in artistic culture must consider cognitive structures, related to the perception of movement, that are both primary and secondary. These pre-motor and motor brain structures make judgments about patterns and can, eventually, have been responsible for the basis of what we now call art. Though it appears static, visual art and sculpture contain movements of all sorts, overt and subtle, to which we unconsciously respond. Such mental responses are clearly adaptive.

Lakoff gives a cognitive hypothesis which suggests that between pre-motor and motor neurology we enable visual schematics such as *on* (above, contact support) and *into* (container, path, source/goal). In fact, the roots of these words have a physical reference. These are primitive mental devices useful for any number of scenarios and over time were utilized along with other behaviors to make art.

If cognition enables art, and if there is nothing more than a synaptic width between them, why can't we say that in our species they evolved together? We have sight for an adaptive reason. We don't have sight *for* cognition. We see to perceive, evaluate, and understand, to recreate so as to avoid or to approach. When those behaviors moved from cognition to durable forms, whether a cave wall or canvas, does that make them any less adaptive? That infinitesimally small margin between what can or cannot be rightfully labelled an adaptation is one of a few evolved mechanisms supposedly separating us from other apes.

As Lakoff suggests, our complex intellectual functions are only possible through our complex visual perceptions. Visual art and form are products of mirror neurons, or more precisely motor neurons, which fire in response to seeing other actions or in response to attempting coordinated action. The artist creatively sees in the finished piece her movements as does the viewer imaginatively. There might be other movements implied in the seeing as well, guesswork implied in theory of mind. On a related note, we often confront with imagined physical vigor various types of visual ambiguity.

Somewhat like Heath, Semir Zeki tackles neurobiology by addressing ambiguity, and he says that art in many ways duplicates brain functions in its attempt to understand an environment. In other words, any variants we find in art simply reflect how our brain tries to accommodate ambiguity we find in the physical world, including the minds of other people. Art can seem ambiguous because our brains are always in an interpretive mode where often there is no single answer. Zeki's point seems to be that any ambiguity we find in art is not special since the brain itself is not a passive receptor but an active meaning maker. Rather than being discarded as useless, ambiguity becomes part of trying to find solutions.

As an example, Zeki offers *The Girl with a Pearl Earring* (c. 1665) by Johannes Vermeer. He says the young female subject is simultaneously inviting, distant, erotic, chaste, resentful, pleased and reflects the various interpretations of a face we would likely confront in real life. The brain, therefore, is an active meaning maker of various information inputs. Or at least the brain tries to make meaningfully coherent the bits and pieces it perceives. What Zeki implies is that there are adaptive functions in how our brains actively process data to make the best survival decisions, and these processes have transferred over to our bracing acceptance of ambiguity in art.

Differing from the ambiguity hypothesis is Robert Pepperell's idea that there is a deliberate dichotomy in art. The viewer is aware of what is represented while simultaneously mindful of how the art was physically constructed. There is conflict in the completeness of what is perceived and in how its materials were manipulated. For example, there could be form and pattern in spite of scattered lines and rough textures. A viewer might ask: What am I seeing and how was this made? Aesthetic experience, says Pepperell, lies in the strength of the dichotomy or how we negotiate expectations and meaning.

While the brain prefers stability in processing its environment it is designed to handle ambiguity and, in fact, invites mental play. Zeki's conclusion is that in terms of neurobiology, ambiguity is not precisely uncertainty but the realization that our consciousness can accommodate a number of solutions to any one problem.

Thomas Scott-Phillips suggests that visual art is not only a form of non-verbal communication but also demonstrates the artist's intent to communicate in an artificially outward presentation. Scott-Phillips says that the viewer grasps the signs and symbols of communication but understands she has to determine the communicator's meaning.

Such problem solving in the face of ambiguity means juggling a number of mental representations at once, along with memory and predictions about the future. There can be, therefore, several solutions or outcomes to a problem. We evolved our brains to anticipate and accept more than one answer in some situations.

Representation and Metarepresentation

Robert Turner and Charles Whitehead go as far as saying that representations can change brain structure. Artistic culture has a significant impact on how one sees herself in relation to others and to the world. In fact, artistic culture, they say, can alter one's consciousness. Therefore, if culture can so affect one's perceptions, and if perceptions are brain functions, then the brain can change via culture. There is a feedback loop between culture and brain patterning. H. Clark Barrett says (perhaps anthropocentrically) that human beings created a cognitive niche where metarepresentations, i.e., highly abstract and symbolic thought or representations about representations, allowed us to improvise in ways that other primates could not. We don't just act or react, but we think about our thoughts in relation to other ideas.

Paleontologist Steven Mithen, in *The Prehistory of the Mind*, in line with evolutionary psychologists, says that our brains come programmed with intelligences and modules formed during our important development in the Pleistocene. Contrary to most psychology of the twentieth century, and against such prominent thinkers as Skinner, Piaget, and even Kohlberg, we are not born with blank minds that get shaped in stages or degrees by our environment. While culture can indeed influence us, we are the culture creators who fashion values, beliefs, and practices from innate faculties and capacities.

Certainly, the physical and social environments have an impact on individuals and groups, but individuals and groups are complicit in creating their environments. The mind is neither a sponge soaking up information nor a computer processing data. Rather, it has an inborn ability to create by imagining, creatively or analytically, outcomes not yet tangible.

According to Howard Gardner there are multiple brain intelligences, such as linguistic, musical, logical, spatial, bodily-kinesthetic, and personal. The brain is not a general processor. Mithen says that *Homo sapiens*, over Neanderthals, were able to connect with fluidity these various brain modules. However, Gardner's intelligences are somewhat general and culturally-determined, whereas Cosmides and Tooby posit faculties that are inherent and pre-loaded with content. Gardner suggests we are programmers; Cosmides and Tooby suggest we are programmed. We are again faced with splitting a hair.

Reconciling the intelligences and faculties is Daniel Sperber and his notion of a module of metarepresentation. Sperber's mega module, says Mithen, acts as a portal that handles both concepts and representations. An integrator. In this way different metarepresentations can bump into and rub off each other. The point is, for Mithen, there is a flexibility of intelligences that permits the human brain by its evolved consciousness to explore itself in order to consider what another mind is thinking.

Perceptions, representations, and thoughts can mingle, combine, and reconfigure in the human brain. I am reluctant to call art behavior, therefore, a byproduct of cognition. At the least, art behavior is a physical product of cognition that helped us solve some problems such as communication, individual and group identity, and mate selection.

Mithen believes that *Homo sapiens* at 100kya was able to bring together specialized intelligences (social, language, natural history, technical, general intelligence) so that by about 33kya our species was able to start making representational and symbolic art and artifacts. Earlier in our discussion I showed how some evolutionists convincingly push this artification date back much further.

In contrast, consider the Neanderthals who did big game hunting but then nothing useful or aesthetic with the bones. The Neanderthal brain was incapable, according to Mithen, of having one domain cross into another. But, also from earlier in our discussion, many thinkers demonstrated that Neanderthals had some forms of what we can term art culture. Many speculate, however, that the Neanderthals died off, perhaps among other reasons, because they did not have the full cognitive flexibility of *Homo sapiens* and so never advanced culturally.

Line or Color?

More specific than representation, Charles Stevens investigates which is more vital in our perceptual brain, line or color. Stevens claims that rather than straight representation, art abstracts selected qualities of what is real. And Stevens believes that since we tend to grasp line drawings quickly, line is how images appear in the brain.

The following few paragraphs, for the most part, summarize Stevens, who explains how part of the visual cortex works. Visual pigment molecules in photoreceptor cells absorb light of one color: red, green, or blue. The pigment molecules create a chemical signal, then sent to neurons, which help the brain understand exactly how blue the sky is on any particular day. How large is this brain region in the cortex? Stevens says it is comparable to a grain of rice or about 10 cubic millimeters, although, it houses approximately 2 million nerve cells. Interestingly, about 90 percent of what we see is generated by the cortex, and is not information landing on the retina.

Also housed in this tiny area are about 20 billion synapses. This means that each nerve cell in the rice-sized region is getting signals from about 10,000 other nerves. This tiny region contains about 40 miles of nerve strands. Neurotransmitters can flood a synapse in a heightened state of pleasure or emotion, similar to what happens in a creative act or viewing. Pertinent to this small visual region, some nerves attend only to signals from the right eye, while others to the left eye. Inputs from both eyes are treated separately so we can form the shape and depth of an object.

Many visual cortex neurons are blind to color. The nerves respond, rather, to edges or lines. When what we perceive as color hits the visual cortex it is treated as a different process. Stevens' point seems to be that line predominates, followed by color. Data from the visual cortex is sent to other brain regions for processing, which then communicate back to the visual cortex influencing the representation.

Stevens presents a handy account of what happens and how signals are processed. For our purposes, though, there is no indication of the adaptive function. Yet it's not difficult to see. Visual cues are cognitively processed to help us negotiate any given environment. We are programmed to respond to visual signals as part of evolved behavior, and at some point began transferring those cognitive skills to the making of art so as to influence the behavior of others.

Seeing Reality Abstractly

For Semir Zeki, in "Neural Concept Formation and Art," any artistic creation or viewing is a brain function. Seeing the outside world of objects, people, and events in abstract outlines is a biological perception for understanding reality. Zeki says that art relies on a basic brain function, forming abstraction. Because the brain is effective in homing in on information and then processing it, ideas about that information are retained. Art derives from the brain's ability and need to take information as abstraction and store or communicate it.

The brain's neurobiology, says Zeki, abstracts so that it is not required to maintain the millions upon millions of pieces of particular data. This claim is similar to Stevens' assertion that art is a process of abstracting what we see. Therefore, art is the physical extension, not a byproduct, of cognition. Detailed information about an object, event, or person is necessary for the making of an abstraction, of course. And realistic representations are memory aids. Importantly, however, an abstraction is a shortcut for the memory, and most likely an adaptive function. Abstraction is not necessarily a higher brain activity, and we are not conscious of the process in visual areas but only the results.

Zeki says that a brain cell in the primary visual cortex sees a straight line but in reality abstracts it, since it knows the code straight-line and initially ignores color or orientation. Apparently, this is also true in brain cells which respond to motion. Ignoring the form or color of an object means abstracting it to focus on where the object is moving or what it is doing. This abstracting process is also true of color, says Zeki. When we see an object, brain cells that respond to color ignore shape or form and compute color only.

Zeki postulates that these abstracting processes are responsible for the brain's making an *ideal* form of something that consists of separate parts, such as shape, color, and motion. The brain formulates a generality with which to work at present or in the future. The brain's evolved efficiency is such that there are quick visual responses that are separate, combined into a general idea that consists of many parts. Zeki suggests that what we call art is an intended result of the brain's ability to make and decipher such abstractions. Art is the physical extension of adapted brain functions, not random fluff.

We see this abstracting process, too, in ambiguity. Our brains enjoy participating in objects, events, or visual perceptions that are not complete. Consider the popularity of jig-saw puzzles. We can see, using our brain, an image that is open to multiple interpretations. The brain processes the parts and attempts to come up with an ideal, or answer, of which there might be several alternative ideal forms. Clearly this is an adaptive function, for the brain will select what it computes as the best solution in what-if-then situations.

Knowledge, Beauty, and Neutrality

In his article, "Artistic Creativity and the Brain," Zeki reminds us that Darwin, in *On the Origin of Species*, reveals how variation, along with competition and inheritance, is a pillar of natural selection. For us, one such variation would be the brain. Although Zeki cannot pinpoint any specific brain structure that is evolving, he does say that variability is clear across populations in terms of intelligence and creativity. Yet, in spite of individual differences or even disabilities, we can communicate universally about art, and while artists differ individually, the same universal neurobiological brain functions help them create.

Variation is a linchpin to natural selection, and one can see from this discussion that material culture and art behaviors for sexual selection or social selection are cognitive adaptations implicated in competition and inheritance. The art and culture making variables helped adapt individuals to each other, to groups, and groups to larger social structures.

Like neuroscientists, artists create to build a knowledge base about how one experiences the physical world. Comprehension of, to the extent that one seeks control over, the physical or social environments is essential to our evolved behaviors. Mondrian, says Zeki, wanted to find a truth about form and found it to be the straight line, much as a neuroscientist will look for how the brain processes a fundamental property of an object or event. For instance, Zeki implies that visual brain cells will quickly process an object or event from various mental perspectives so as to compose a concept that can be applied to different scenarios in the future.

Hideaki Kawabta and Zeki tested people to identify paintings of any subject they considered beautiful, neutral, or ugly. The people were then asked to view those paintings while in an fMRI scanner. The researchers determined that distinct categories of perception are located in specific brain areas. The question, then, is whether beauty is in the object (Plato) or subject (Kant). Overlooked is how some aspects of beauty are evolved aesthetics, such as facial shapes. Some brain areas do not act in isolation but are active in any viewing so that there is reciprocity between areas. We see and evaluate with our brains.

Kawabta and Zeki go on to say that the anterior cingulate and the left parietal cortices seem particularly implicated in someone distinguishing between beauty and neutrality. The anterior cingulate is large and associated with positive and pleasurable emotions, both of which were prominent in the contrast of beautiful versus neutral. What this means is that an aesthetic experience is seated in an emotional brain state. However, they say fMRI does not reveal invisible areas that most likely are also implicated.

Worth noting is that although an fMRI can read which areas of the brain light up as active, the machine cannot measure human outcome. How we physically and emotionally register stimuli might carry some universals, but how we respond can be more personal than cultural.

Kawabta and Zeki conclude that there is specific brain activation in response to a color or motion stimulus, but meanwhile what we call beauty operates on a continuous wave that has approval or disapproval added to it based on a number of factors. This wave is variable so that an individual could change her mind about the merits of a painting. Both beauty and ugliness activate the visual motion center of the brain, which suggests that one's inflection of activity is what creates the value judgment. The motor cortex activates in socially inappropriate behavior. Fearful images or threatening voices and anger determine whether one should retreat or act upon the stimulus.

From Discontinuity to Essence

In his paper "The Disunity of Consciousness," Zeki adds to our discussion by saying that what we refer to singularly as consciousness is incorrect. There are manifold, hierarchical consciousnesses across the brain. We just happen to think of ourselves as a singular self perceiving the world. What Zeki means is that, as Stevens showed, visual systems in the brain are spread out in different areas. The processing of an image does not automatically constitute consciousness of the image. These findings have vast implications concerning the immense variability of creating and responding to art culture, from any one individual to groups.

Zeki says that the hierarchy runs something like this: location, color, motion, orientation. There is asynchrony because of comparisons dealing with, first, color, and then, motion, among the evaluation of other visual characteristics. These percepts are not necessarily bound in a straight line. Rather they might bind *after* consciousness, so that there is a tier of micro-consciousness, macro-consciousness, and then a unified consciousness. We might not be aware of the micro and macro operations, only the unified one, which equates to the perceiving self.

Obviously, this process of perception serves an adaptive function to singularize moving threats or opportunities. Additionally, from the cultural side, this process of perception enables any artificer to play with the viewer's conscious and subconscious mind and manipulate it toward conformity.

Here's a slightly different organizing process. V.S. Ramachandran and William Hirstein, in "The Science of Art," claim that the deep structures of aesthetic experience cross individual and cultural differences. They suggest, therefore, that there is a genetic mechanism underlying the appreciation of art. They see visual art not as mere depiction of reality but "to enhance, transcend, or indeed even to *distort* reality." According to these authors, the Hindu word *rasa*, flavor or emotion (via the Sanskrit word for sap or essence) captures what the artist attempts – to get to the quintessence of the object or event. Ramachandran and Hirstein believe that an artist is trying to make for the viewer an experience of intensity and vitality less through the direct copying of an image and more to elemental qualities. The evolved brain itself does this by abstracting pertinent information and discarding what is unnecessary.

Ramachandran and Hirstein say that studies in rats have demonstrated there may be neurons in the brain that represent certain effects, patterns, shapes, or forms, and artists can intensify or *stimulate* those neurons. In this way art might dramatically distort form but yet somehow in such distortion effectively capture and accentuate the *rasa* of whatever is depicted. I hesitate to offer any Eurocentric, modern example, but think of works by Post-Impressionists. The authors suggest that there is a parallel in music, which operates on prehistoric or even primate vocalizations related to emotional responses programmed in our brains.

We seem to find satisfaction in being able to recognize patterns, being able to group disparate marks into an image, just as we respond positively and favorably to color combinations in, for example, bunched flowers. We can call these response activities, say Ramachandran and Hirstein, primitive forms. In any given moment we can attend to and compute only so much informational input. The demands of attention place stress on our limbic system, which signals the cortex, which sends information back to the limbic system. We need to know how to react when we see an object or event. Such a fury of activity across regions will create some partiality or shortcuts in what is perceived.

Brain Sight and Insight

In "Visual Art and the Visual Brain," Zeki's Woodhull Lecture, the simple claim is that we see with our brains through the primary visual cortex V1 area. Afterwards, the brain takes those visual signals and sends them to other parts of the brain for processing, e.g., in terms of interpreting form, color, and motion. Consider how all brains are similarly built on a general plan, but not all brains have the same neural connections.

Visual brain cells are particular, responding initially to part of the field of vision, called the receptive field, and later to bits within that receptive field. Clearly, this is adaptive, since we cannot instantaneously process all pieces of visual data that come our way and must selectively focus on objects and events related to food resources and similar attention-grabbers. But here, too, not every synaptic connection in every brain has developed equally.

In his lecture, Zeki also covers facial recognition. Portraiture broadcasts universal emotions and personality modules while yet capturing a distinct character. Zeki says there is, rather than one aesthetic, a range which includes an aesthetic for color, movements, forms, and expressions which all integrate into what we call the aesthetic experience. Keep in mind that these aesthetic predilections evolved over long periods of time. We might go as far as asking why we even see. Mice and moles have very low vision and survive fine.

We evolved sight and visual brain areas to comprehend fully our physical and social environments. In some way, seeing is metaphorically grasping, which is control. We adapted visual and cognitive networks to advance our values and beliefs. Nevertheless, in spite of vast differences among cultural groups, as a species we can recognize positively what another culture finds beautiful. We might link, therefore, brain sight and insight, especially in how we evaluate the behavior of others. As Andrea Kantrowitz says, our knowledge is not only perception and movement but also proprioception or the sense of our body in the world. Artwork helps us introspect about our physical presence in reality in relation to other people, real or imagined.

One fourth of the brain is dedicated to vision, since there is important tactile and locational information about objects, events, or people that is most efficiently acquired and processed visually. The world we inhabit is constantly moving and changing, and objects and events can appear differently from different perspectives. The brain, however, will visually try to stabilize such objects and events to obtain a good reading. Clearly our active participation in compiling bits of data into what we see and process is a survival mechanism. We try to understand what we see.

A leaf in summer is green whether in sunshine, shade, or darkness, and the brain seeks constancy in constructing this essential color based on prior knowledge and experience. Art is a way for us to transfer what the brain sees in its process of assembling an idea of an object or event. For Zeki, visual art is an attempt to get to the concepts of objects and events, a function of awareness that served us in evolution. Of course any categorization or classification of discrete data into concept maps implies evaluation.

While specific details matter, Zeki's thrust in his Woodhull Lecture is that the brain compiles details only to find a more essential design. Part of our cognitive success derives from taking particles of information and connecting them, to have at hand an idea of something to compare with another version. Since we are highly social creatures, we can do this with people as well. Clearly this sorting and compiling capacity could have adaptive functions when confronting a stranger or entering an unknown area. On one viewing of visual art many brain cells respond, but then fewer and fewer with repeated viewings, even if invariant, indicating that an idea of the object or event has been stored for future reference.

For example, there are brain cells that selectively respond to orientation, e.g., in the appearance of lines. These receptor cells, says Zeki, are important building components in the perception of form, intuited by Kasimir Malevich and Mondrian. In a similar fashion, Kandinsky focused on the square and the rectangle. There are cells in certain regions that respond specifically to certain shapes, lines and movements, colors or backgrounds, while some cells respond to spots or bars. Color is a brain perception used for comparison: eat the riper, more colorful fruit. Willem de Kooning, says Zeki, underscores color without form.

What at first seems incomprehensible is, on second viewing, a cognitive challenge, part of our evolved legacy in dealing with an intensely colorful world bursting with patterns and shapes. We make art to think; art makes us think. We evolved to make sense of our thoughts and sensations.

Beauty and Cognitive Emotions

G. Gabrielle Starr, in *Feeling Beauty*, asks some seminal questions about how emotions are related to an aesthetic experience, how individual experiences vary, and what type of knowledge aesthetic experience offers us. Key ideas Starr brings to the discussion include the nature of beauty, our vocabulary in interpreting and explaining what we like, and the restless continuum of aesthetic experience that at once can underscore and yet alter culture.

Starr says that the brain's so-called default mode network is implicated in aesthetic experience. This network operates when we are not focused on an external task but rather are engaged on inner thoughts, such as daydreaming, recalling memories, or planning for the future. The network has also been implicated in theory of mind and creativity since both involve visual signals assuming no strenuous cognitive activity is occurring. Importantly, says Starr, the default mode helps us imagine our own sense of self, which impacts any aesthetic experience. In other words, what we call an aesthetic experience has less to do with an external object or event and more with our inner states of mind and emotions in making value judgments. See, too, Tone Roald.

The take home point here is that anything aesthetic is a combination of, Starr rightly notes, what is "both thought and felt; it is something 'cognitive,' 'sensory,' and 'emotional'" (16). In this way, a work of art can alter how we view ourselves, see the world, or interact with other people. There is cognitive play and pretense as well as uncertainty, an organic process of working out details, designing a pattern, and making an evaluation. These functions are adaptive since they help us interact or avoid situations, people, or groups. But Starr is more to the cognitive and less to the evolutionary, as has been outlined in most of this primer, saying that aesthetic rewards are of greater importance than those in resources for survival or sexual selection.

From an evolutionary perspective, Gianluca Consoli finds art arising specifically from emotions. Initially, there were individual and fixed primary emotions shared with other primates, such as fear, anger, disgust, joy, and sadness. These gave way to shared, self-conscious, and creative secondary emotions related to cognition, such as planning. A primary emotion is what we first feel; later, a secondary emotion can be a mixture of primary and thinking about such primary emotions. It is from the secondary emotions that we have art. Very early art moved primary emotions to secondary. What this suggests is that our aesthetic emotions are tied evolutionarily to our primary emotions.

Randy Thornhill talks about the affective nature of beauty and insists that our feelings for beauty are adaptations that would have helped us find the best foods, the right mates, and the most comfortable habitats. Olaf Breidbach would say that beauty is a signal of the "biological reality" of an object. Consider, for instance, our evolved facial recognition mechanisms which draw us to a healthy physique and clear eyes, or how we are repulsed by the sight of rotten food or bodily sores.

Starr's point is that artistic culture jolts our perceptions and emotions, causing a re-boot of our internal system. Granted, evolutionary psychology can focus on the adapted brain to the exclusion of the individual. But the arguments previously outlined by people such as Mithen, Miller, Cosmides, and Tooby hold more sway on the adaptive function of the arts than simply saying adapted mechanisms are by-products of cognition. Cognition and reason are themselves part of the adapted mind.

However, Starr is right to suggest that theory of mind can help explain why we have artistic culture. For instance, we could be reading the mind of someone in a painting. There is someone absorbed in an act; one contemplating a question, such as a marriage proposal; one engaged in a dilemma of some sort not disclosed. If the representations in the painting are too challenging (e.g., surreal art) we consider the mind of the artist. With abstract art we focus on reading our own minds.

But Starr fails to see that evolutionary psychology looks to basic impulses. No one (e.g., Dissanayake, Pagel, or Mesoudi) would presume to say the specifics of any culture (i.e., the particular expression of such impulses) evidence direct evolutionary impacts. One of the reasons for her displeasure with evolutionary psychology, understandably, is that there is little accommodation for individual differences and individual experience. Starr ponders that neural emotions responding to art are not necessarily the same as those we encounter on any given day. But that means that art has no evolutionary value at all, since emotional reaction is our default response.

Starr is right to suggest than any stimulation of a pleasure or reward center in the brain is not automatically equivalent to an aesthetic evaluation. Certainly, as our brains evolved, with the expansion of the neocortex and the integration of longer synapses, any sense of value judgment probably came later. Nonetheless, evolution builds on what is already there, so responses related to survival and reproduction, especially sexual selection and mate choice, could be responsible for reward responses to some visual (aesthetic) stimuli. This is not to be absolute in terms of evolution and adaptation, since epigenetics and cultural groups exert a strong influence on what we view and how we evaluate beauty.

At any rate, one can agree with Starr that an aesthetic experience is emotional and cognitive, both of which in turn rely on memories. There are social implications to artistic culture since an aesthetic experience enlivens how we feel, perceive, and relate to others. In a sense, what we call aesthetics helps us establish and recalibrate certain values so that one can imagine oneself acting or moving in special or different ways. The neurobiology of aesthetics is not completely cerebral but includes motor neurons that help us move physically and emotionally in relation to the art's creator, our own bodies, and in association with the outside world.

Ritual Art

On a final note, let us turn to Ara Norenzayan and Will Gervais who have written on the cultural evolution of religion. This is related to our subject since many religions include ritual arts and cultural social practices. The sociality of religion is premised on the belief that one helps another at a cost to himself. Essentially all religions emphasize a ritualistic social unity. As opposed to individual or sexual selection, here we have a notion of group selection.

Norenzayan and Gervais say that some religious beliefs are adaptations for cooperative behavior to benefit the individual in a group. Clearly, though, religion is a group benefit, as if the individuals act and think with one mind. The authors go on to say that religious art is an extension of

cognitive functions such as theory of mind, where one spec-
ulates about the thinking of another, even an imagined or
supernatural person. Theory of mind gave rise to a realization that there was an arena separate from reality, something
non-physical. This is counterintuitive but part of our history, and such mentalizing produced its own distinct set of
representations.

Here, too, we see a strong group component, since shared
beliefs both spread within the group and unified its members. With an acceptance of supernatural beings, moral
codes could be enforced without human intervention. And
finally, also on a group selection level, shared beliefs, values, and practices helped establish rules to punish members
and to compete with other groups. All of these practices
could have been, and to some extent still are, achieved visually through symbolic forms. I cannot think of any religion
that does not have symbols or icons. In fact, in our history,
many groups in competition have destroyed and still strive
to eradicate each other's religious symbols and artifacts.

The authors conducted research on these premises and
found that, generally speaking, there is distrust of atheists.
Why? Because atheists have no symbols, no signs, no art.
This might not be entirely true. Some atheists focus on images from nature or other abstract forms, and there is their
atomic whirl. But surely in a large degree almost all religions have a variety of artistic culture, for display or wor-
ship, and often related to rituals. If one is tied to the artistic
culture of a religion and its rituals, then one is part of the
group. Such devote participation, say Norenzayan and Gervais, is difficult to pretend. Nevertheless, there are many
atheists, the fourth largest group after Christians, Muslims,
and Hindus. So, religion is probably cultural and not genetic. At base, though, as demonstrated, culture is genetic. And
there is wide cultural variation.

"Wherever art displays itself, there would seem to be an absence of truth." Quintilian

"Art remains the one way possible of speaking truth." Robert Browning

—

"Art is the right hand of nature." Schiller

CONCLUSION

Let me conclude with a question that has been implied all along. Can there be consilience among the arts and sciences?

According to the team of Robert and Michele Root-Bernstein, a basic assumption is that the arts and sciences are so separate that any individual is either an artist or a scientist but not both. Their research is important for our discussion since it validates how the evolved brain has an adapted mind capable of mental and creative flexibility. In other words, the arts and sciences share many tendencies and behaviors in common, part of our evolutionary heritage in imaginatively solving problems related to fitness, survival, and reproduction. Such misguided separation of art and science goes back to C.P. Snow and his essay, "The Two Cultures."

Snow, say the Root-Bernsteins, asserts that those in the arts and those in the sciences work and think differently and cannot communicate with each other. In fact, Snow goes as far as saying that these divisive groups do not even share the same emotions. Of course, from an evolutionary perspective, or from work by a psychologist such as Paul Ekman, such assertions are ridiculous. We all evolved from the same ancestors and are different only in terms of individual genes and cultural influences.

The Root-Bernsteins argue that Snow, in spite of the proliferation of his idea, is wrong based on their research. Certainly the results of what an artist or a scientist does can differ, and each might have a different set of goals. Nevertheless, even Zeki sees the arts as a knowledge-gathering and sharing mechanism. Be that as it may, the Root-Bernsteins say that how artists and scientists work is creatively similar.

Since the human brain evolved, we all share a similarly evolved mind. Our limbic brain works with our neocortex. And though different, the left hemisphere communicates with the right. Jerome Kagan, among other developmental psychologists, has proved that most people have innate temperamental differences, and the Root-Bernsteins have found that many scientists are artists and many artists are scientists. Variation is a part of natural selection, so individual differences do not eradicate our generally evolved behaviors.

Contrary to Snow, there are not two separate cultures but one that includes polymaths capable of various types of ingenious creativity to address or solve problems.

The Root-Bernsteins assert that inventive scientists share the profile of an artist and often make art. Statistically, about one fifth of Nobel laureates in literature are avid naturalists and prone to admiring Darwinian thought. What the Root-Bernsteins have discovered from their research is that "creative people *integrate* apparently disparate skills, talents, and activities into a synergistic whole." Highly sophisticated creative minds apparently blend on a functional level both the artistic and the cognitive.

From previous notes in this small book, we know there is contiguity between emotions and cognition. Artistic creativity can be an attempt to express emotions cognitively or to express cognition emotionally. Sometimes art is simply an emotional presentation with little thought.

The Root-Bernsteins cite Howard Gardner, who as we know developed the idea of multiple intelligences and wrote, among other books, *Creating Minds*. Gardner suggests that creativity takes place in cognitive domains, such as the visual-spatial, musical, kinesthetic, or logico-mathematical. Yet Gardner, say the Root-Bernsteins, doubts that there is any horizontality among faculties which would permit cross domain creativity. Why such strict divisions?

CONCLUSION

The Root-Bernsteins, however, have uncovered a plethora of polymaths in the sciences and arts, which means that faculties or intelligence domains interact often and with meaningful results. A creative personality can be either or both artistic and scientific. Steven Mithen made a similar claim by saying that our cognitive fluidity is what helped us survive over other hominin species, most recently, the Neanderthals.

To be fair, the Root-Bernsteins say that Gardner most likely emphasizes the results or finished products of either an artist or a scientist and not the process that overreaches any such product. Gardner's *frames of mind*, for instance, are based on specific, quantifiable outcomes. The error, according to the Root-Bernsteins, is that Gardner has neglected the means by which creators, whether scientists or artists, think through the process of their work.

For our purposes, the conclusions to be drawn are that the polymath mind demonstrates how evolution rigged the brain to adapt creatively in solving problems. This is not a late development but goes back in some early formation at least to *Homo habilis*. There is no either/or. As a species we are both scientifically and artistically inclined. We should acknowledge how these two perspectives often work in tandem, not at odds. We *seem* to be scientists and use the scientific method to solve tangible problems. But we are also artists who solve complex social and interpersonal problems. Both of these approaches are cognitively and emotionally related adaptations that have helped us survive.

Let us hope that our inclination for the arts and sciences will work to continue human flourishing for all people without compromising the survival of other species or planet earth.

"The highest problem of every art is, by means of appearances, to produce the illusion of a loftier reality." Goethe

"Art is nothing but the highest sagacity and exertion of human nature; and what nature will he honor who honors not the human?" Lavater

"In no circumstances whatever can man be comfortable without art." Ruskin

BIBLIOGRAPHY

As a primer, I've tried to keep this book simple. But the literature on art and evolution is vast. Readers are encouraged to investigate materials by the authors listed and the works they cite. Some material on this updated yet selective list might not be cited in the text above.

Aiken, Nancy. 1998. *The Biological Origins of Art*. Westport, CT: Praeger.

—. 2013."Aesthetics and Evolution." *Aisthesis. Pratiche, linguaggi esaperi dell'estetico* 6.2. 61-73.

Alexander, Richard. 1990. *How Did Humans Evolve? Reflections on the Uniquely Unique Species*. Ann Arbor, MI: U Michigan.

Al-Shawf, Laith, et al. 2015. "Human Emotions: An Evolutionary Psychological Perspective." *Emotion Review*.

Alvard, Michael S. 2003. "The Adaptive Nature of Culture." *Evolutionary Anthropology* 12: 136-149.

Armstrong, Paul B. 2013. *How Literature Plays with the Brain: The Neuroscience of Reading and Art*. Baltimore, MD: Johns Hopkins UP.

Atran, Scott. 2010. "The Evolution of Religion: How Cognitive By-Products, Adaptive Learning Heuristics, Ritual Displays, and Group Competition Generate Deep Commitments to Prosocial Religions." *Biological Theory* 5.1. 18-30.

Ayala, Francisco and Camilo J. Cela-Conde. 2017. *Processes in Human Evolution*. Oxford: OUP.

Balter, Michael. 2009. "On the Origin of Art and Symbolism." *Science* 323: 709-711.

Barrett, H. Clark. 2007. "The Hominid Entry into the Cognitive Niche." *The Evolution of Mind*. Steven W. Gangestad and Jeffry A. Simpson, eds. NY: Guilford.

Barrett, Louise. 2011. *Beyond the Brain: How Body and Environment Shape Animal and Human Minds*. Princeton: PUP.

Bartels, Andreas and Semir Zeki. 2005. "Brain dynamics during natural viewing conditions – A new guide for mapping connectivity in vivo." *NeuroImage* 24. 339-349.

Baumeister, Roy F. 2005. *The Cultural Animal: Human Nature, Meaning, and Social Life*. Oxford: OUP.

Bednarik, Robert G. 2013. "Pleistocene Palaeoart Art of Asia." *Arts* 2. 46-76.

Boyd, Brian. 2009. "Art and Selection." *Philosophy and Literature* 33. 2-4-220.

Brandt, Per Aage. 2006. "Form and Meaning in Art." *The Artful Mind: Cognitive Science and the Riddle of Human Creativity*. Mark Turner, ed. Oxford: OUP. 171-188.

Breidbach, Olaf. 2003. "The Beauties and the Beautiful – Some Considerations from the Perspective of Neuronal Aesthetics." *Evolutionary Aesthetics*. Eckart Voland and Karl Grammer, eds. Berlin: Springer. 39-68.

Brown, Jerram L. 1975. *The Evolution of Behavior*. NY: Norton.

Buckner, Randy L. et al. 2008. "The Brain's Default Network: Anatomy, Function, and Relevance to Disease." *Annals of the New York Academy of Sciences* 1124. 1-38.

Buss, David M. 2008. *Evolutionary Psychology: The New Science of the Mind*. Third edition. Boston: Pearson.

Carroll, Joseph. 2013. "The Imagined World." Response to Symposium Question: "Why make art?" *Island: Ideas, Writing, Culture*.

Carruthers, Peter and Andrew Chamberlain. 2000. Introduction. *Evolution and the Human Mind*. P. Carruthers and A. Chamberlain, eds. Cambridge: CUP. 1-12.

Castro, Laureano and Miguel A. Toro. 2004. "The Evolution of Culture: From Primate Social Learning to Human Culture." *PNAS* 101.27. 10235-10240.

Chalmers, David J. 2003. "Consciousness and its Place in Nature." *Blackwell Guide to Philosophy of Mind*. Stephen Stich and Ted A. Warfield, eds. Blackwell. 102-142.

Charlesworth, Brian and Deborah. 2003. *Evolution: A Very Short Introduction*. Oxford: OUP.

Chase, Philip. 1999. "Symbolism as Reference and Symbolism as Culture." *The Evolution of Culture*. Robin Dunbar, Chris Knight, Camilla Powers, eds. New Brunswick: Rutgers UP. 34-49.

Chauvet, Jean-Marie, Eliette Brunel Deschamps, Christian Hillaire. 1996. *Dawn of Art: The Chauvet Cave*. NY: Harry N. Abrams.

Cheney, Dorothy L. and Robert M. Seyfarth. 1990. *How Monkeys See the World: Inside the Mind of Another Species*. Chicago: U Chicago P.

Clark, Andy and David Chalmers. 1998. "The Extended Mind." *Analysis* 58.1. 7-19.

Clottes, Jean. 2011. *What Is Paleolithic Art?* Translated by Oliver Y. Martin and Robert D. Martin. Chicago: U Chicago P, 2016.

—. 2013. "Why Did They Draw in Those Caves?" *Time and Mind: The Journal of Archaeology, Consciousness and Culture* 6.1. 7-14.

Coe, Kathryn. 1992. "Art: The Replicable Unit – An Inquiry Into the Possible Origin of Art as a Social Behavior." *Journal of Social and Evolutionary Systems* 15.2. 217-234.

—. 2003. *The Ancestress Hypothesis: Visual Art as Adaptation*. New Brunswick, NJ: Rutgers UP.

—. 2013. "Can Science Lead Us to a Definition of Art?" *Aisthesis Pratiche, linguaggi e saperi dell'estetico*. 6.2. 153-177.

Conkey, Margaret W. 2001. "Hunting for Images, Gathering up Meanings: Art for Life in Hunting-Gathering Societies." *Hunter-Gatherers: An Interdisciplinary Perspective*. Catherine Painter-Brick, Robert H. Layton, Peter Rowley-Conwy, eds. Cambridge: CUP. 267-291.

Consoli, Gianluca. 2015. "Early Art and the Evolution of Grounded Emotions." *Aisthesis. Pratiche, linguaggi esaperi dell'estetico*. 8.1. 147-156.

Cosmides, Leda and John Tooby. 1992. "The Psychological Foundations of Culture." *The Adapted Mind: Evolutionary Psychology and the Generation of Culture*. Jerome Barkow, et al., eds. NY: OUP, 1992. 19-136.

Coss, Richard G. 2003. "The Role of Evolved Perceptual Biases in Art and Design." *Evolutionary Aesthetics*. Eckart Voland and Karl Grammer, eds. Berlin: Springer. 69-130.

Damasio, Antonio. 1999. *The Feeling of What Happens: Body and Emotion in the Making of Consciousness*. Orlando: Harvest.

Darwin, Charles. 1859. *On the Origin of Species*. Joseph Carroll, ed. Ontario, CN: Broadview, 2003.

—. *The Descent of Man and Selection in Relation to Sex*. 1871. 1879 (second edition cited here). London: Penguin Books, 2004.

—. 1872. *The Expression of the Emotions in Man and Animals*. London: Penguin, 2009.

Davies, Stephen. 2012. *The Artful Species: Aesthetics, Art, and Evolution*. Oxford: OUP.

Deacon, Terrence W. 1997. *The Symbolic Species: The Co-Evolution of Language and the Brain*. NY: Norton, 1997.

—. 2006. "The Aesthetic Faculty." *The Artful Mind: Cognitive Science and the Riddle of Human Creativity*. Mark Turner, ed. Oxford: OUP. 21-53.

De Smedt, Johan and Helen De Cruz. 2012. "Human Artistic Behaviour: Adaptation, Byproduct, or Cultural Group Selection?" *Philosophy of Behavioral Biology*. Kathryn S. Plaisance and Thomas Reydon, eds. Springer. 167-187.

De Sousa, Ronald. 2004. "Is Art an Adaptation? Prospects for an Evolutionary Perspective on Beauty." *The Journal of Aesthetics and Art Criticism* 62.2. 109-118.

Deter-Wolf, Aaron, et al. 2016. "The World's Oldest Tattoos." *Journal of Archaeological Science: Reports* 5. 19-24.

de Waal, Frans. 1996. *Good Natured: The Origin of Right and Wrong in Humans and Other Animals*. Cambridge, MA: Harvard UP.

—. 1999. "Anthropomorphism and Anthropodenial: Consistency in our Thinking about Humans and Other Animals." *Philosophical Topics* 27.1. 255-280.

—. 2006. With Robert Wright, Christine M. Korsgaard, Philip Kitcher, and Peter Singer. *Primates and Philosophers: How Morality Evolved*. Stephen Macedo, ed. Princeton: PUP.

—. 2007 (1982). *Chimpanzee Politics: Power and Sex Among Apes*. Baltimore, MD: Johns Hopkins UP.

—. 2009. *The Age of Empathy: Nature's Lessons For a Kinder Society*. NY: Harmony Books.

—. 2013. *The Bonobo and the Atheist: In Search of Humanism Among the Primates*. NY: Norton.

Dissanayake, Ellen. 1995. *Homo Aestheticus: Where Art Comes From and Why*. Seattle: U Washington P.

—. 1998. "Komar and Melamid Discover Pleistocene Taste." *Philosophy and Literature* 22.2. 486-496.

—. 2001. "Aesthetica Incunabula." *Philosophy and Literature* 25. 335-346.

—. 2001. "Becoming *Homo Aestheticus*: Sources of Aesthetic Imagination in Mother-Infant Interactions." *SubStance* 94/95. 85-103.

Donald, Merlin. 2006. "Art and Cognitive Evolution." *The Artful Mind: Cognitive Science and the Riddle of Human Creativity*. Mark Turner, ed. Oxford: OUP. 3-20.

—. 2013. "Mimesis Theory Re-Examined, Twenty Years after the Fact." *Evolution of Mind, Brain, and Culture*. Gary Hatfield and Holy Pittman, eds. Philadelphia: U Pennsylvania Museum of Archaeology and Anthropology. 169-192.

Dutton, Denis. 2009. *The Art Instinct: Beauty, Pleasure, and Human Evolution*. NY: Bloomsbury.

Dunbar, Robin. 1998. "The Social Brain Hypothesis." *Evolutionary Anthropology* 6(5): 178-190.

Eccles, John C. 1992. "Evolution of Consciousness." *Proceedings of the National Academy of Sciences* 89. 7320-7324.

—. 2014. *Human Evolution: A Pelican Introduction*. London: Penguin.

Ekman, Paul and Richard J. Davidson, eds. 1994. *The Nature of Emotion*. NY: Oxford UP.

Epstein, Russell. 2004. "Consciousness, Art, and the Brain: Lessons from Marcel Proust." *Consciousness and Cognition* 13. 213-240.

Feinberg, Todd E. and Jon M. Mallat. 2016. *The Ancient Origins of Consciousness: How the Brain Created Experience*. Cambridge, MA: MIT Press.

BIBLIOGRAPHY

Fischer, John. 2009. "Art Styles as Cultural Cognitive Maps." *American Anthropologist* 63.1. 79-93.

Fossey, Dian. 1983. *Gorillas in the Mist*. Boston: Houghton Mifflin, 2000.

Gallese, Vittorio and David Freedberg. 2007. "Mirror and Canonical Neurons are Crucial Elements in Esthetic Response." *Trends in Cognitive Science* 11.10. 411.

Gardner, Howard. 1993. *Creating Minds*. NY: Basic Books.

—. 2011. *Frames of Mind: The Theory of Multiple Intelligences*. Third edition. NY: Basic Books.

Galdikas, Biruté and Nancy Briggs. 1999. *Orangutan Odyssey*. NY: Harry N. Abrams.

Goh, Joshua O. and Denise C. Park. 2009. "Culture Sculpts the Perceptual Brain." *Cultural Neuroscience: Cultural Influences on Brain Function*. Joan Chiao, ed. Amsterdam: Elsevier. 95-111.

Gombrich, E.H. 1978. *The Story of Art*. Oxford: Phaidon.

Goodall, Jane. 1986 *The Chimpanzees of Gombe: Patterns of Behavior*. Cambridge: Harvard UP.

Gottesman, Sarah. 2016. "The Neuroscience of Op Art." *Artsy*.

Gray, Martin Paul. 2010. "Cave Art and the Evolution of the Human Mind." M.A. Thesis. Victoria Univ. of Wellington. Advisor: Kim Sterelny.

Guthrie, R. Dale. 2005. *The Nature of Paleolithic Art*. Chicago: U Chicago P.

Haidle, Miriam N. 2014. "Examining the Evolution of Artistic Capacities: Searching for Mushrooms?" *Art as Behaviour: An Ethological Approach to Visual and Verbal Art, Music an Architecture*. Christa Sütterlin, et al., eds. Oldenberg: Bis-Verlag der Carl von Ossietzky Universität Oldenberg. 237-251.

Heath, Shirley Brice. 2006. "Dynamics of Completion." *The Artful Mind: Cognitive Science and the Riddle of Human Creativity*. Mark Turner, ed. Oxford: OUP. 133-150.

Hodgson, Derek. 2006. "Altered States of Consciousness and Palaeoart: An Alternative Neurovisual Explanation." *Cambridge Archaeological Journal* 16.1. 27-37.

—. 2006. "Understanding the Origins of Paleoart: The Neurovisual Resonance Theory ad Brain Functioning." *PaleoAnthropology*. 54-67.

Hogan, Patrick Colm. 2010. "On Being Moved: Cognition and Emotion in Literature and Film." *Introduction to Cognitive Cultural Studies*. Lisa Zunshine, ed. Baltimore: Johns Hopkins UP. 237-256.

Humphrey, Nicholas. 1983. *Consciousness Regained: Chapters in the Development of Mind*. Oxford: OUP.

Ishizu, Tomohiro and Semir Zeki. 2011. "Toward a Brain-Based Theory of Beauty." *Plos One* 6.7. e21852. 1-10.

Jablonka, Eva and Marion J. Lamb. 2005. *Evolution in Four Dimensions: Genetic, Epigenetic, Behavioral, and Symbolic Variation in the History of Life*. Cambridge, MA: MIT/Bradford.

Janson, H.W. 1991. *History of Art*. Fourth edition, revised by Anthony F. Janson. NY: Harry N. Abrams.

Junker, Thomas. 2010. "Art as a Biological Adaptation, or: Why Modern Humans Replaced the Neanderthals." *Quartär* 57. 171-178.

Kagan, Jerome. 2010. *The Temperamental Thread: How Genes, Culture, Time, and Luck Make Us Who We Are*. NY: Dana P.

Kandel, Eric R. 2012. *The Age of Insight: The Quest to Understand the Unconscious in Art, Mind, and Brain, from Vienna 1900 to the Present*. NY: Random House.

Kantrowitz, Andrea. 2012. "The Man Behind the Curtain: What Cognitive Science Reveals about Drawing." *The Journal of Aesthetic Education* 46.1. 1-14.

Kardan, Omid et al. 2015. "Neighborhood Greenspace and Health in a Large Urban Center." *Scientific Reports* 5.11610.

Kawabta, Hideaki, and Semir Zeki. 2004. "Neural Correlates of Beauty." *Journal of Neurophysiology* 91. 1699-1705.

Kennedy, Henry and Kenneth Knoblauch. 2005. "Imagery, Art and Biological Aspects of Visual Consciousness." *Word and Image* 21.2. 124-135.

Klein, Richard G. 2009. *The Human Career: Human Biological and Cultural Origins*. Third edition. Chicago: U Chicago P.

Knight, Chris, Camilla Power and Ian Watts. 1995. "The Human Symbolic Revolution: A Darwinian Account." *Cambridge Archaeological Journal*. 75-114.

Koch, Christof. 2012. *Consciousness: Confessions of a Romantic Reductionist*. Cambridge, MA: MIT P.

Köhler, Wolfgang. 1925. *The Mentality of Apes*. Ella Winter, trans. London: Routledge and Kegan Paul, 1973.

Kohn, Marek and Steven Mithen. 1999. "Handaxes: Products of Sexual Selection?" *Antiquity* 73: 518-26.

Kuhn, Steven L. and Mary C. Stiner. 1998. "Middle Paleolithic 'Creativity.'" *Creativity in Human Evolution and Prehistory*. Steven Mithen, ed. London: Routledge. 143-164.

Lakoff, George. 2006. "The Neuroscience of Form in Art." *The Artful Mind: Cognitive Science and the Riddle of Human Creativity*. Mark Turner, ed. Oxford: OUP. 153-169.

Layton, Robert. 1985. "The Cultural Context of Hunter-Gatherer Rock Art." *Man* 20.3. 434-453.

BIBLIOGRAPHY

LeDoux, Joseph. 1996. *The Emotional Brain: The Mysterious Underpinnings of Emotional Life.* NY: Simon and Schuster.

Lewis-Williams, J.D. 1986. "Cognitive and Optical Illusions in San Rock Art Research." *Current Anthropology* 27.2. 171-178.

—. 1994. "Rock Art and Ritual: Southern Africa and Beyond." *Complutum* 5. 277-289.

—. 1997. "Harnessing the Brain: Vision and Shamanism in Upper Paleolithic Western Europe." *Beyond Art: Pleistocene Image and Symbol.* Margaret Conkey, Olga Soffer, Deborah Stratmann, Nina G. Jablonski, eds. San Francisco: Memoirs of the California Academy of Sciences, number 23. 321-342.

—. 2013. *San Rock Art.* Athens, Ohio: Ohio UP.

Livingstone, Margaret. 2002. *Vision and Art: The Biology of Seeing.* NY: Abrams.

Lock, Anthony. 2015. "Evolutionary Aesthetics, the Interrelationship Between Viewer and Artist, and New Zealandism." *ASEBL Journal* 11.2. 3-36.

Lycett, Stephen J. 2011. "'Most Beautiful and Most Wonderful': Those Endless Stone Tool Forms." *Journal of Evolutionary Psychology* 9.2: 143-171.

Mayr, Ernst. 1991. *One Long Argument: Charles Darwin and the Genesis of Modern Evolutionary Thought.* Cambridge: Harvard UP.

McBrearty, Sally and Alison S. Brooks. 2000. "The Revolution that Wasn't: A New Interpretation of the Origin of Modern Human Behavior." *Journal of Human Evolution* 39: 453-563.

McGrew, William C. 1998. "Culture in Nonhuman Primates?" *Annual Review of Anthropology.* 27. 301-328.

Mesoudi, Alex. 2011. *Cultural Evolution: How Darwinian Theory Can Explain Human Culture and Synthesize the Social Sciences.* Chicago: U Chicago P.

Miller, Geoffrey. 2001. *The Mating Mind: How Sexual Choice Shaped the Evolution of Human Nature.* NY: Anchor Books.

Mithen, Steven. 1996. *The Prehistory of the Mind: The Cognitive Origins of Art, Religion and Science.* NY: Thames and Hudson.

—. 2009. "The Music Instinct: The Evolutionary Basis of Musicality." *The Neurosciences and Music III – Disorders and Plasticity: Annals of the N.Y. Academy of Sciences* 1169. 3-12.

Mohanan K.P. 2011. "The Biological Foundations of Art: Denis Dutton's Art Instinct." *Journal of Genetics* 90.3. 1-5.

Moulton, Samuel T. and Stephen M. Kosslyn. 2009. "Imagining Predictions: Mental Imagery as Mental Emulation." *Philosophical Transactions of The Royal Society B* 364. 1273-1280.

Noë, Alva. 2015. *Strange Tools: Art and Human Nature*. NY: Hill and Wang.

Norenzayan, Ara and Will M. Gervais. 2011. "The Cultural Evolution of Religion." *Creating Consilience: Integrating the Sciences and the Humanities*. Edward Slingerland and Mark Collard, eds. Oxford: OUP. 243-265.

Nowell, April. 2013. "Cognition, Behavioral Modernity, and the Archaeological Record of the Middle and Early Upper Paleolithic." *Evolution of Mind, Brain, and Culture*. Philadelphia: U of Pennsylvania Museum of Archaeology and Anthropology. 235-262.

—. 2021. *Growing Up in the Ice Age*. Oxford: Oxbow Books.

Osvath, Mathias and Peter Gärdenfors. 2005. "Oldowan Culture and the Evolution of Anticipatory Cognition." *Lund University Cognitive Science* 122.

Pagel, Mark. 2012. *Wired for Culture: Origins of the Human Social Mind*. NY: Norton.

Parker, Sue Taylor and Kathleen Rita Gibson, eds. 1990. *"Language" and Intelligence in Monkeys and Apes: Comparative Developmental Perspectives*. Cambridge: Cambridge UP.

Pepperell, Robert. 2015. "Artworks as Dichotomous Objects: Implications for the Scientific Study of Aesthetic Experience." *Frontiers in Human Neuroscience* 9:295.

Pinker, Steven. 1997. *How the Mind Works*. NY: Norton.

Portera, Mariagrazia. 2013. "Evolutionary Aesthetics. A non-human approach to the human appreciation of beauty." Poster. V Congress of the Italian Society for Evolutionary Biology. Trento.

Prum, Richard O. 2012. "Aesthetic Evolution by Mate Choice: Darwin's *Really* Dangerous Idea." *Philosophical Transactions of the Royal Society B* 367. 2253-2265.

—. 2013. "Coevolutionary Aesthetics in Human and Biotic Artworlds." *Biology and Philosophy*.

—. 2017. *The Evolution of Beauty*. NY: Doubleday.

Rajković, Milan and Miloš Milovanović. 14 June 2015. "The Artists Who Forged Themselves: Detecting Creativity in Art." Unpublished.

Ramachandran, V.S. and William Hirstein. 1999. "The Science of Art: A Neurological Theory of Aesthetic Experience." *Journal of Consciousness Studies* 6.6/7. 15-51.

Richerson, Peter J. and Robert Boyd. 2005. *Not by Genes Alone: How Culture Transformed Human Evolution*. Chicago: U Chicago P.

Roald, Tone. 2015. *The Subject of Aesthetics: A Psychology of Art and Experience*. Leiden: Brill.

Root-Bernstein, Robert and Michele Root-Bernstein. 2004. "Artistic Scientists and Scientific Artists: The link between polymathy and crea-

BIBLIOGRAPHY

tivity." *Creativity: From Potential to Realization.* Robert J. Sternberg, Elena L. Grigorenko, Jerome L. Singer, eds. Wash., DC: American Psychological Association. 127-151.

Rule, Nicholas O. 2014. "Cultural Neuroscience: A Historical Introduction and Overview." *Online Readings in Psychology and Culture* 9.2.

Runciman, W.G. 2005. "Culture Does Evolve." *History and Theory* 44. 1-13.

Scott-Phillips, Thomas C. "What is Art? A Pragmatic Perspective." *Think* 40, 14. 87-91.

Seeley, W.P. 2013. "Art, Meaning, and Reception: A Question of Methods for a Cognitive Neuroscience of Art." *British Journal of Aesthetics* 53.4. 443-460.

Seghers, Eveline. 2014. "Cross-Species Comparison in the Evolutionary Study of Art: A Cognitive Approach to the Ape Art Debate." *Review of General Psychology* 18.4. 263-272.

Shryock, Andrew and Daniel Lord Smail, eds. 2011. *Deep History: The Architecture of Past and Present.* Berkeley: U California P.

Shryock, Andrew, Thomas R. Trautmann, Clive Gamble. 2011. "Imagining the Human in Deep Time." Shryock, Andrew and Daniel Lord Smail, eds. *Deep History: The Architecture of Past and Present.* Berkeley: U California P. 21-52.

Smail, Daniel Lord and Andrew Shryock. 2011. "Body." Shryock, Andrew and Daniel Lord Smail, eds. *Deep History: The Architecture of Past and Present.* Berkeley: U California P. 55-77.

Smail, Daniel Lord, Mary C. Stiner, Timothy Earle. 2011. "Goods." Shryock, Andrew and Daniel Lord Smail, eds. *Deep History: The Architecture of Past and Present.* Berkeley: U California P. 219-241.

Snow, C.P. 1959. "The Two Cultures." *The Two Cultures.* Stefan Collini, ed. Cambridge: CUP, 1998.

Sparshott, Francis. 1997. "Art and Anthropology." *The Journal of Aesthetics and Art Criticism* 55.3. 239-243.

Spolsky, Ellen. 2010. "Making 'Quite Anew': Brain Modularity and Creativity." *Introduction to Cognitive Cultural Studies.* Lisa Zunshine, ed. Baltimore: Johns Hopkins UP. 84-102.

Starr, G. Gabrielle. 2013. *Feeling Beauty: The Neuroscience of Aesthetic Experience.* Cambridge, MA: MIT P.

Steen, Francis. 2006. "A Cognitive Account of Aesthetics." *The Artful Mind: Cognitive Science and the Riddle of Human Creativity.* Mark Turner, ed. Oxford: OUP. 57-71.

Stevens, Charles F. 2001. "Line versus color: the brain and the language of visual arts." *The Origins of Creativity.* Karl Pfenninger and Valerie R. Shubik, editors. Oxford: OUP. 177-189.

Stokstad, Marilyn and Michael W. Cothren. 2011. *Art History*. Fourth edition. Boston: Prentice Hall.

Straffon, Larissa Mendoza. 2014. *Art in the Making: The Evolutionary Origins of Visual Art as a Communication Signal*. Leiden: U of Leiden.

—. 2016. "Signalling in Style: On Cooperation, Identity and the Origins of Visual Art." *Understanding Cultural Traits*. Fabrizio Panebianco and Emanuele Serrelli, eds. Switzerland: Springer. 357-373.

Teske, Joanna Klara. 2011. "Should Not 'Beauty and Pleasure' be Complemented with 'Cognition and 'Communication'? Reflections Upon Denis Dutton's *Art Instinct: Beauty, Pleasure and Human Evolution*." *Diametros* 28. 105-114.

Texier, Pierre-Jean et al. 2010. "A Howiesons Poort Tradition of Engraving Ostrich Eggshell Containers Dated to 60,000 Years Ago at Diepkloof Rock Shelter, South Africa." *PNAS* 107.14. 6180-6185.

Thornhill, Randy. 2003. "Darwinian Aesthetics Informs Traditional Aesthetics." *Evolutionary Aesthetics*. Eckart Voland and Karl Grammer, eds. Berlin: Springer. 9-35.

Tomasello, Michael. 2014. *A Natural History of Human Thinking*. Cambridge, MA: Harvard UP.

Tooby, John and Leda Cosmides. 2001. "Does Beauty Build Adapted Minds? Toward an Evolutionary Theory of Aesthetics, Fiction and the Arts." *SubStance* 94/95. 6-27.

Turner, Mark, ed. 2006. *The Artful Mind: Cognitive Science and the Riddle of Human Creativity*. Oxford: OUP.

—. 2006. "The Art of Compression." *The Artful Mind: Cognitive Science and the Riddle of Human Creativity*. Mark Turner, ed. Oxford: OUP. 93-113.

Turner, Robert and Charles Whitehead. 2008. "How Collective Representations Can Change the Structure of the Brain." *Journal of Consciousness Studies* 15.10.11. 43-57.

Verpooten, Jan and Mark Nelissen. 2010. "Sensory Exploitation and Cultural Transmission: The Late Emergence of Iconic Representation in Human Evolution." *Theory in Biosciences* 129 (2-3). 211-221.

Voland, Eckart and Karl Grammer, eds. 2003. *Evolutionary Aesthetics*. Berlin: Springer.

von Petzinger, Genevieve. 2016. *The First Signs: Unlocking the Mysteries of the World's Oldest Symbols*. NY: Atria.

von Petzinger, Genevieve, April Nowell. 2014. "A Place in Time: Situating Chauvet within the Long Chronology of Symbolic Behavioral Development." *Journal of Human Evolution* 74. 37-54.

Williams, George C. 1966. *Adaptation and Natural Selection*. Princeton: PUP.

BIBLIOGRAPHY

Wilson, Edward O. 1999. *Consilience: The Unity of Knowledge*. NY: Vintage.

Zahavi, Amotz. 1975. "Mate Selection – A Selection for a Handicap." *Journal of Theoretical Biology* 53. 205-214.

Zeki, Semir. 1997. "Visual Art and the Visual Brain." The Woodhull Lecture 1995. *Proceedings of the Royal Institution of Great Britain* 68. 29-63.

—. 2001. "Artistic Creativity and the Brain." *Science* 293.5527. 51-52.

—. 2002. "Neural Concept Formation and Art: Dante, Michelangelo, Wagner." *Journal of Consciousness Studies* 9.3. 53-76.

—. 2003. "The Disunity of Consciousness." *Trends in Cognitive Science* 7.5. 214-218.

—. 2006. "The Neurobiology of Ambiguity." *The Artful Mind: Cognitive Science and the Riddle of Human Creativity*. Mark Turner, ed. Oxford: OUP. 243-270.

Zunshine, Lisa. 2012. *Getting Inside Your Head: What Cognitive Science Can Tell Us About Popular Culture*. Baltimore, MD: The Johns Hopkins University Press.

—, ed. 2010. *Introduction to Cognitive Cultural Studies*. Baltimore: Johns Hopkins UP.

ART AND ADAPTATION

—

"Art requires philosophy, just as philosophy requires art. Otherwise, what would become of beauty?" Gaugin

"The aim of art is not to represent the outward appearance of things but their inward significance." Aristotle

INDEX

INDEX

INDEX

ABOUT BIBLIOTEKOS

Bibliotekos was started in 2009 with the aim of publishing five themed literary anthologies of contemporary writers of fiction, creative nonfiction, and poetry. The following anthologies include, on a highly selective basis, writers from all over the world. *Pain and Memory: Reflections on the Strength of the Human Spirit in Suffering* (2009); *Common Boundary: Stories of Immigration* (2010); *Battle Runes: Writings on War* (2011); *Being Human: Call of the Wild* (2012); *Patterns of Faith and Puzzles of Doubt* (2013). A literary website, noted below, is kept active and up-to-date with author profiles, reviews, and some original writing.

In 2019 a sister site was created, *Literary Veganism: An Online Journal*, which publishes creative writing by, for, and about vegans, animal rights, and environmental ethics. See the link below. The ASEBL website is devoted to analytical prose on animal and environmental ethics.

Bibliotekos website: www.ebibliotekos.com

Literary Veganism: www.litvegan.net

ASEBL website: www.asebl.net

AUTHOR BIO

Scholar, author, and editor Gregory F. Tague, Ph.D. (1998 New York University) is Professor Emeritus of English and Interdisciplinary Studies (1998-2023) and was the founder and senior developer of The Evolutionary Studies Collaborative at St. Francis College, N.Y. He also initiated a number of Darwin-inspired Moral Sense Colloquia and other multidisciplinary events. In addition to many short-form publications, books include: *Forest Sovereignty: Wildlife Sustainability and Ethics* (Oxford: Peter Lang 2025); *The Vegan Evolution: Transforming Diets and Agriculture* (London: Routledge 2022); *An Ape Ethic and the Question of Personhood* (Lanham: Lexington Books 2020); *Art and Adaptability: Consciousness and Cognitive Culture* (Leiden: Brill|Rodopi 2018); *Evolution and Human Culture* (Leiden: Brill|Rodopi 2016); and *Making Mind: Moral Sense and Consciousness in Philosophy, Science, and Literature* (Amsterdam: Rodopi 2014). While Tague's published work spans a number of disciplines, his current interests hover around environmental and animal ethics. Tague has also written or edited nine other academic books or literary anthologies, including *Character and Consciousness* (2005), *Origins of English Dramatic Modernism* (2010), and *Puzzles of Faith and Patterns of Doubt* (2013). Tague is the founding editor of the *ASEBL* website, and is general editor of the Bibliotekos literary site and *Literary Veganism: An Online Journal.*

For more information about Professor Tague, including details about all of his books, visit
https://sites.google.com/site/gftague/

NOTES